ELECTRI

by Paul Rosenberg

This book is the property of:

Tracy Horner

~~Robert Spang~~

Please call and return if found

Telephone: (   )   -

Additional copies of this book and
many others are always available from:

MAINTENANCE TROUBLESHOOTING

273 Polly Drummond Road
Newark, Delaware 19711
TOLL FREE: (800) 755-7672

"The biggest name in little books."

First Edition

Engineering Publications Group—EPG, Inc.
374 Circle of Progress
Pottstown, PA 19464-3800

Library of Congress Cataloging-in-Publication Data

Rosenberg, Paul A.
    Electrical Pal/by Paul Rosenberg,—First Edition
    Engineering Publications Group—EPG, Inc.
    ISBN 0-9652171-0-8

    1. Electrical engineering—Handbooks, manuals, etc.
96-84636

Foreword

For several years I have been asked to create a compendium of tables, charts, graphs, diagrams and terminology relating to the field of electricity. The Electrical Pal is my first endeavor to create a pocket guide for the electrical marketplace that will serve as an all encompassing publication in an easy-to-use format. I have tried to use a design and typeface that will hold the most information per page while at the same time making a particular topic quick to reference.

Information covered in this manual is necessary for anyone in the field to have in one's possession at all times. Naturally, a topic may have been overlooked or not discussed in depth to suit all tradespeople. I will constantly monitor and update this book on a regular basis to not only include requested additional material, but to add new material from the ever growing amount of high technology as it develops.

Paul Rosenberg

How To Use This Book

For your convenience in using this book, we have taken the table of contents heading and expanded it to include the sequential sub-headings of charts, tables, graphs, etc.

Since there is no index to this book, AND NO SEQUENTIAL PAGE NUMBERS, all sub-headings will be referenced at the bottom center of the page they are located on as specified in the expanded table of contents. For example, in Chapter one, the third sub-heading is a chart on AC/DC Formulas. That chart will be found on page 1-3.

CHAPTER 1
ELECTRICAL FORMULAS

CHAPTER 2
ELECTRONICS AND ELECTRONIC SYMBOLS

CHAPTER 3
RACEWAYS & WIRING

VII

CHAPTER 4
COMMUNICATIONS

CHAPTER 5
FIBER OPTICS

CHAPTER 6
MOTORS

CHAPTER 7
TRANSFORMERS

CHAPTER 8
PLAN SYMBOLS

CHAPTER 9
LIGHTING

CHAPTER 10
CONVERSION FACTORS AND UNITS
OF MEASUREMENT

CHAPTER 11
MATERIALS & TOOLS

CHAPTER 12
TOLL FREE NUMBERS FOR ELECTRICAL SUPPLIERS

CHAPTER 13
GLOSSARY

NOTES

CHAPTER 1
ELECTRICAL FORMULAS

CALCULATING ROOT-MEAN-SQUARE		
To Convert	**To**	**Multiply By**
rms	Average	.9
rms	Peak	1.414
Average	rms	1.111
Average	Peak	1.567
Peak	rms	.707
Peak	Average	.937
Peak	Peak-to-peak	2

POWER FORMULAS — 1ϕ, 3ϕ		
Phase	**To Find**	**Use Formula**
1ϕ	I	$I = \dfrac{VA}{V}$
1ϕ	VA	$VA = I \times V$
1ϕ	V	$V = \dfrac{VA}{I}$
3ϕ	I	$I = \dfrac{VA}{V \times \sqrt{3}}$
3ϕ	VA	$VA = I \times V \times \sqrt{3}$

PANELBOARDS

SINGLE-PHASE — 3-WIRE SYSTEMS

40 A	100 A	150 A	225 A	400 A
70 A	125 A	200 A	300 A	600 A

THREE-PHASE — 4-WIRE SYSTEMS

60 A	150 A	225 A	400 A
125 A	200 A	300 A	600 A

GUTTERS and WIREWAYS

2-1/2" x 2-1/2"	6" x 6"	10" x 10"
4" x 4"	8" x 8"	

These sizes are available in 12", 24", 36", 48", and 60" lengths.

DISCONNECTS

30 A	200 A	800 A	1600 A
60 A	400 A	1200 A	1800 A
100 A	600 A	1400 A	

SIZES OF PULL BOXES and JUNCTION BOXES

4" x 4" x 4"	10" x 8" x 4"	12" x 12" x 6"
6" x 4" x 4"	10" x 8" x 6"	12" x 12" x 8"
6" x 6" x 4"	10" x 10" x 4"	15" x 12" x 4"
6" x 6" x 6"	10" x 10" x 6"	15" x 12" x 6"
8" x 6" x 4"	10" x 10" x 8"	18" x 12" x 4"
8" x 6" x 6"	12" x 8" x 4"	18" x 12" x 6"
8" x 6" x 8"	12" x 8" x 6"	18" x 18" x 4"
8" x 8" x 4"	12" x 10" x 4"	18" x 18" x 6"
8" x 8" x 6"	12" x 10" x 6"	24" x 18" x 6"
8" x 8" x 8"	12" x 12" x 4"	24" x 24" x 6"
		24" x 24" x 8"

AC/DC FORMULAS

TO FIND	DIRECT CURRENT	ALTERNATING CURRENT 1φ, 115, or 120 V	ALTERNATING CURRENT 1φ, 208, 230, or 240 V	ALTERNATING CURRENT 3φ – ALL VOLTAGES
AMPERES WHEN HORSEPOWER IS KNOWN	$\dfrac{HP \times 746}{E \times E_{FF}}$	$\dfrac{HP \times 746}{E \times E_{FF} \times PF}$	$\dfrac{HP \times 746}{E \times E_{FF} \times PF}$	$\dfrac{HP \times 746}{1.73 \times E \times E_{FF} \times PF}$
AMPERES WHEN KILOWATTS IS KNOWN	$\dfrac{kW \times 1000}{E}$	$\dfrac{kW \times 1000}{E \times PF}$	$\dfrac{kW \times 1000}{E \times PF}$	$\dfrac{kW \times 1000}{1.73 \times E \times PF}$
AMPERES WHEN KVA IS KNOWN		$\dfrac{kVA \times 1000}{E}$	$\dfrac{kVA \times 1000}{E}$	$\dfrac{kVA \times 1000}{1.73 \times E}$
KILOWATTS	$\dfrac{I \times E}{1000}$	$\dfrac{I \times E \times PF}{1000}$	$\dfrac{I \times E \times PF}{1000}$	$\dfrac{I \times E \times 1.73 \times PF}{1000}$
KILOVOLT-AMPS		$\dfrac{I \times E}{1000}$	$\dfrac{I \times E}{1000}$	$\dfrac{I \times E \times 1.73}{1000}$
HORSEPOWER (OUTPUT)	$\dfrac{I \times E \times E_{FF}}{746}$	$\dfrac{I \times E \times E_{FF} \times PF}{746}$	$\dfrac{I \times E \times E_{FF} \times PF}{746}$	$\dfrac{I \times E \times 1.73 \times E_{FF} \times PF}{746}$

BUSWAY or BUSDUCT	
1 φ	3 φ
225 A	225 A
400 A	400 A
600 A	600 A
800 A	800 A
1000 A	1000 A
1200 A	1200 A
1350 A	1350 A
1600 A	1600 A
2000 A	2000 A
2500 A	2500 A
3000 A	3000 A
4000 A	4000 A
5000 A	5000 A

CBs and FUSES

15	70	225	800
20	80	250	1000
25	90	300	1200
30	100	350	1600
35	110	400	2000
40	125	450	2500
45	150	500	3000
50	175	600	4000
60	200	700	5000
			6000

For fuses only, additional standard sizes are 1, 3, 6, and 10.

SWITCHBOARDS or SWITCHGEARS	
1 φ	3 φ
200 A	400 A
400 A	600 A
600 A	800 A
800 A	1200 A
1200 A	1600 A
1600 A	2000 A
2000 A	2500 A
2500 A	3000 A
3000 A	4000 A
4000 A	

OHM'S LAW/POWER FORMULAS

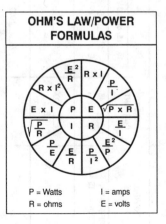

P = Watts   I = amps
R = ohms   E = volts

POWER FACTOR

An ac electrical system carries two types of power: (1) true power, watts, that pulls the load (Note: Mechanical load reflects back into an ac system as resistance.) and (2) reactive power, vars, that generates magnetism within inductive equipment. The vector sum of these two will give actual volt-amperes flowing in the circuit (see diagram right). Power factor is the cosine of the angle between true power and volt-amperes.

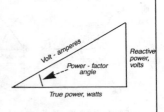

SINGLE-PHASE POWER

Power of a single-phase ac circuit equals voltage times current times power factor: $P_{watts} = E_{volts} 1_{amps} pf$.

To figure reactive power, vars squared equal volt-amperes squared minus power squared, or $vars^2 = v\text{-}a(1\text{-}pr^2)$.

THREE-PHASE AC CIRCUITS

Three-phase is the most common polyphase system. It's a system having three distinct voltages out of step with one another. There are 120° between each voltage. At any instant the algebraic sum (measured up and down from centerline) of these three voltages is zero. When one voltage is zero, the other two are 86.6% maximum and have opposite signs.

The three phases are generated by placing each phase coil in the generator 120° apart, mechanically. Rotating dc field will then cut each phase coil in turn, inducing voltage in each out of step with the other two.

UTILIZATION OF THREE-PHASE POWER

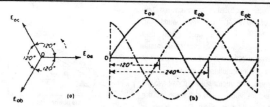

(a)   (b)

Sine waves above are actually an oscil-
lograph trace taken at any point in a three-
phase system. (Each voltage or current
wave actually comes from a separate wire
but are shown for comparison on common
base.) Big advantage of three phase comes
in motor application when fed to stator windings (six poles) is like a three-cylinder
instead of a one-cylinder (single-phase)
motor.

Y-CONNECTION

Consider the three windings as primary
of transformer. Current in all windings
equals line current, but volts across windings = 0.577 x line volts.

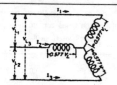

DELTA CONNECTION

Winding voltages equal line voltages,
but currents split up so 0.577 I_{line} flows
through windings. Transformers operate at
5% capacity with one winding open.

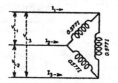

FOUR-WIRE SYSTEM

Most popular secondary distribution
setup. V_1 is usually 208v which feeds small
power loads. Lighting loads at 120v tap from
any line to neutral.

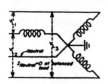

HORSEPOWER FORMULAS

TO FIND	USE FORMULA
HP	$HP = \dfrac{E \times I \times E_{FF}}{746}$
I	$I = \dfrac{HP \times 746}{E \times E_{FF} \times PF}$

HP = horsepower
E = volts (Note: A "V" after a value also
     indicates volts – for example, 240 V.)
E_{FF} = efficiency
746 = watts per horsepower
PF = power factor

VOLTAGE DROP FORMULAS — 1φ, 3φ

PHASE	TO FIND	USE FORMULA
1φ	VD	$VD = \dfrac{2 \times R \times L \times I}{CM}$
3φ	VD	$VD = \dfrac{2 \times R \times L \times I}{CM}$

DECIBELS vs. VOLT & POWER RATIOS

Voltage	Power	+ DB -	Power	Voltage
1.000	1.000	0.0	1.000	1.000
1.059	1.122	0.5	0.891	0.944
1.122	1.259	1.0	0.794	0.891
1.189	1.413	1.5	0.708	0.847
1.259	1.585	2.0	0.631	0.794
1.334	1.778	2.5	0.562	0.750
1.413	1.995	3.0	0.501	0.708
1.496	2.239	3.5	0.447	0.668
1.585	2.512	4.0	0.398	0.631
1.679	2.818	4.5	0.355	0.596
1.778	3.162	5.0	0.316	0.562
1.884	3.548	5.5	0.282	0.531
1.995	3.981	6.0	0.251	0.501
2.113	4.467	6.5	0.224	0.473
2.239	5.012	7.0	0.200	0.447
2.371	5.623	7.5	0.178	0.422
2.512	6.310	8.0	0.158	0.398
2.661	7.079	8.5	0.141	0.376
2.818	7.943	9.0	0.126	0.355
2.985	8.913	9.5	0.112	0.335
3.162	10.000	10.0	0.100	0.316
3.350	11.220	10.5	0.089	0.299
3.548	12.589	11.0	0.079	0.282
3.758	14.125	11.5	0.071	0.266
3.981	15.849	12.0	0.063	0.251
4.217	17.783	12.5	0.056	0.237
4.467	19.953	13.0	0.050	0.224
4.732	22.387	13.5	0.045	0.211
5.012	25.119	14.0	0.040	0.200
5.309	28.184	14.5	0.035	0.188
5.623	31.623	15.0	0.032	0.178
5.957	35.481	15.5	0.028	0.168
6.310	39.811	16.0	0.025	0.158
6.683	44.668	16.5	0.022	0.150
7.079	50.119	17.0	0.020	0.141
7.499	56.234	17.5	0.018	0.133
7.943	63.096	18.0	0.016	0.126
8.414	70.795	18.5	0.014	0.119
8.913	79.433	19.0	0.013	0.112
9.441	89.125	19.5	0.011	0.106
10.0	100	20.0	0.010	0.100
31.6	1000	30.0	0.001	0.0316
100.0	10000	40.0	0.0001	0.01
316.2	10^5	50.0	0.00001	0.00316
1000	10^6	60.0	10^{-6}	0.001
3162	10^7	70.0	10^{-7}	0.000316
10000	10^8	80.0	10^{-8}	0.001
31620	10^9	90.0	10^{-9}	0.0000316
10^5	10^{10}	100.0	10^{-10}	10^{-5}
316200	10^{11}	110.0	10^{-11}	0.00000316
10^6	10^{12}	120.0	10^{-12}	10^{-6}

FORMULAS FOR ELECTRICITY

(1) Ohms Law (DC Current):

$$\text{Current in amps} = \frac{\text{Voltage in volts}}{\text{Resistance in ohms}} = \frac{\text{Power in watts}}{\text{Voltage in volts}}$$

$$\text{Current in amps} = \sqrt{\frac{\text{Power in watts}}{\text{Resistance in ohms}}}$$

Voltage in volts = Current in amps x Resistance in ohms

Voltage in volts = Power in watts / Current in amps

$$\text{Voltage in volts} = \sqrt{\text{Power in watts} \times \text{Resistance in ohms}}$$

Power in watts = (Current in amps)^2 x Resistance in ohms

Power in watts = Voltage in volts x Current in amps

Power in watts = (Voltage in volts)^2 / Resistance in Ohms

Resistance in ohms = Voltage in volts / Current in amps

Resistance in ohms = Power in watts / (Current in amps)^2

(2) Resistors in Series (values in Ohms):

Total Resistance = Resistance$_1$ + Resistance$_2$ + ...Resistance$_n$

(3) Two Resistors in Parallel (values in Ohms):

$$\text{Total Resistance} = \frac{\text{Resistance}_1 \times \text{Resistance}_2}{\text{Resistance}_1 + \text{Resistance}_2}$$

(4) Multiple Resistors in Parallel (values in Ohms):

$$\text{Total Resistance} = \frac{1}{1 / \text{Resistance}_1 + \text{Resistance}_2 + ...\text{Resistance}_n}$$

FORMULAS FOR ELECTRICITY

(5) Ohms Law (AC Current):

In the following AC Ohms Law formulas, θ is the phase angle in degrees by which current lags voltage (in inductive circuit) or by which current leads voltage (in a capacitive circuit). In a resonant circuit (such as normal household 120VAC) the phase angle is 0° and Impedance = Resistance

$$\text{Current in amps} = \frac{\text{Voltage in volts}}{\text{Impedance in ohms}}$$

$$\text{Current in amps} = \sqrt{\frac{\text{Power in watts}}{\text{Impedance in ohms} \times \cos\theta}}$$

$$\text{Current in amps} = \frac{\text{Power in watts}}{\text{Voltage in volts} \times \cos\theta}$$

$$\text{Voltage in volts} = \text{Current in amps} \times \text{Impedance in ohms}$$

$$\text{Voltage in volts} = \frac{\text{Power in watts}}{\text{Current in amps} \times \cos\theta}$$

$$\text{Voltage in volts} = \sqrt{\frac{\text{Power in watts} \times \text{Impedance ohm}}{\cos\theta}}$$

$$\text{Impedance in ohms} = \text{Voltage in volts} / \text{Current in amps}$$

$$\text{Impedance in ohms} = \text{Power in watts} / (\text{Current amps}^2 \times \cos\theta)$$

$$\text{Impedance in ohms} = (\text{Voltage in volts}^2 \times \cos\theta) / \text{Power in watts}$$

$$\text{Power in watts} = \text{Current in amps}^2 \times \text{Impedance in ohms} \times \cos\theta$$

$$\text{Power in watts} = \text{Current in amps} \times \text{Voltage in volts} \times \cos\theta$$

$$\text{Power in watts} = \frac{(\text{Voltage in volts})^2 \times \cos\theta}{\text{Impedance in ohms}}$$

FORMULAS FOR ELECTRICITY

(6) Resonance: - f

Resonant frequency in hertz (where $X_L = X_C$) =

$$\frac{1}{2\pi \sqrt{\text{Indicates in henrys} \times \text{Capacitance in farads}}}$$

(7) Reactance: - X

Reactance in ohms of an inductance is X_L
Reactance in ohms of a capacitance is X_C

$X_L = 2\pi(\text{frequency in hertz} \times \text{Inductance in henrys})$
$X_C = 1 / (2\pi(\text{frequency in hertz} \times \text{Capacitance in farads})$

(8) Impedance: - Z

Impedance in ohms = $\sqrt{\text{Resistance in ohms}_2 + (X_L - X_C)^2}$
(series)

Impedance in ohms = $\dfrac{\text{Resistance in ohms} + \text{Reactance}}{\sqrt{\text{Resistance in ohms}^2 + \text{Reactance}^2}}$
(parallel)

(9) Susceptance: - B

Susceptance in mhos = $\dfrac{\text{Reactance in ohms}}{\text{Resistance in ohms}^2 + \text{Reactance in ohms}^2}$

(10) Admittance: - Y

Admittance in mhos = $\dfrac{\text{Reactance in ohms}}{\sqrt{\text{Resistance in ohms}^2 \times \text{Reactance in ohms}^2}}$

Admittance in mhos = 1 / Impedance in ohms

FORMULAS FOR ELECTRICITY

(11) Power Factor: - pf

Power Factor = cos (Phase Angle)
Power Factor = True Power / Apparent Power
Power Factor = Power in watts / (volts x current in amps)
Power Factor = Resistance in ohms / Impedance in ohms

(12) Q or Figure of Merit: - Q

Q = Inductive Reactance in ohms / Series Resistance in ohms
Q = Capacitive Reactance in ohms / Series Resistance in ohms

(13) Efficiency of any Device:

Efficiency = Output / Input

(14) Sine Wave Voltage and Current:

Effective (RMS) value = 0.707 x Peak value
Effective (RMS) value = 1.11 x Average value
Average value = 0.637 x Peak value
Average value = 0.9 x Effective (RMS) value
Average value = 1.414 x Effective (RMS) value
Peak value = 1.57 x Average value

(15) Decibels: - db

db = 10 Log 10 (power in Watts #1 / Power in Watts #2)
db = 10 Log 10 (Power Ratio)
db = 20 Log 10 (Volts or Amps #1 / Volts or Amps #2)
db = 20 Log 10 (Volts or Current Ratio)
Power Ratio = $10^{(db/10)}$
Voltage or Current ration = $10^{(db/10)}$

If impedances are not equal:

$$db = 20 \log_{10} [(Volt1 \sqrt{Z_2}) / (Volt2 \sqrt{Z_1})]$$

FORMULAS FOR ELECTRICITY

(16) Capacitors in Parallel (values In any farad):

Total Capacitance = Capacitance$_1$ + Capacitance$_2$ + ...Capacitance$_n$

(17) Two Capacitors in Serial (values in any farad):

$$\text{Total Capacitance} = \frac{\text{Capacitance}_1 \times \text{Capacitance}_2}{\text{Capacitance}_1 + \text{Capacitance}_2}$$

(18) Multiple Capacitors in Series (values in any farad):

$$\text{Total Capacitance} = \frac{1}{\text{Capacitance}_1 + \text{Capacitance}_2 + ...\text{Capacitance}_n}$$

(19) Quantity of Electricity in a Capacitor: -Q

Q in coulombs = Capacitance in farads x Volts

(20) Capacitance of a Capacitor: -C

Capacitance in picofarads =

$$0.0885 \times \frac{\text{Dielectric constant} \times \text{area in cm}^2 \times (\text{\# of plates - 1})}{\text{thickness of dielectric in cm}}$$

(21) Self Inductance:

Use the same formulas as those for Resistance, substituting inductance for resistance. When including the effects of coupling, add 2 x mutual inductance if fields are adding and subtract 2 x mutual inductance if the fields are opposing. e.g.

Series: $L_t = L_1 + L_2 + 2M$ or $L_t = L_1 + L_2 - 2M$

Parallel: $L_t = 1 / [(1/L_1 + M) + (1/L_2 + M)]$

FORMULAS FOR ELECTRICITY

(22) Frequency and Wavelength: f and λ

Frequency in kilohertz = (300,000) / wavelength in meters
Frequency in megahertz = (300) / wavelength in meters
Frequency in megahertz = (984) / wavelength in feet
Wavelength in meters = (300,000) / frequency in kilohertz
Wavelength in meters = (300) / frequency in megahertz
Wavelength in meters = (984) / frequency in megahertz

(23) Length of and Antenna:

Quarter-wave antenna:
     Length in feet = 234 / frequency in megahertz

Half-wave antenna:
     Length in feet = 468 / frequency in megahertz

(24) LCR Series Time Circuits:

Time in seconds =
     Inductance in henrys / Resistance in ohms

Time in seconds =
     Capacitance in farads x Resistance in ohms

(25) 70 Volt Loud Speaker Matching Transformer:

Transformer Primary Impedance =
     (Amplifier output volts)^2 / Speaker Power

(26) Time Duration of One Cycle:

10 megahertz	=	100 nanoseconds cycle
4 megahertz	=	250 nanoseconds cycle
1 megahertz	=	1 microsecond cycle
250 kilohertz	=	100 microsecond cycle
100 kilohertz	=	10 microsecond cycle

CAPACITORS

Connected in Series		Connected in Parallel	Connected in Series/Parallel
Two Capacitors	**Three or More Capacitors**		
$C_T = \dfrac{C_1 \times C_2}{C_1 + C_2}$	$\dfrac{1}{C_T} = \dfrac{1}{C_1} + \dfrac{1}{C_2} + ...$	$C_T = C_1 + C_2 + ...$	1. Calculate the capacitance of the parallel branch. $C_T = C_1 + C_2 + ...$
where			2. Calculate the capacitance the series combination.
C_T = total capacitance (in μF)			$C_T = \dfrac{C_1 \times C_2}{C_1 + C_2}$
C_1 = capacitance of capacitor 1 (in μF)			
C_2 = capacitance of capacitor 2 (in μF)			

HORSEPOWER

Current and Voltage Known	Speed and Torque Known
$HP = \dfrac{E \times I \times E_{ff}}{746}$	$HP = \dfrac{rpm \times T}{5252}$
where	where
HP = horsepower	HP = horsepower
E = voltage (volts)	rpm = revolutions per minute
I = current (amps)	T = torque (lb-ft)
E_{ff} = efficiency	

TEMPERATURE CONVERSIONS

Covert °C to °F	Covert °F to °C
°F = (1.8 × °C) + 32	$°C = \dfrac{(°F-32)}{1.8}$

MOTOR TORQUE

Torque	Starting Torque	Nominal Torque Rating
$T = \dfrac{HP \times 5252}{rpm}$	$T = \dfrac{HP \times 5252 \times \%}{rpm}$	$T = \dfrac{HP \times 63,000}{rpm}$
where	where	where
T = torque	HP = horsepower	T = nominal torque rating (in lb-in)
HP = horsepower	5252 = constant $\left(\dfrac{33,000 \text{ lb-ft}}{\pi \times 2} = 5252\right)$	$63,000$ = constant
5252 = constant $\left(\dfrac{33,000 \text{ lb-ft}}{\pi \times 2} = 5252\right)$	rpm = revolutions per minute	HP = horsepower
rpm = revolutions per minute	% = motor class percentage	rpm = revolutions per minute

NOTES

CHAPTER 2
ELECTRONICS &
ELECTRONIC SYMBOLS

RESISTOR COLOR CODES

Color	1st Digit (A)	2nd Digit (B)	Multiplier (C)	Tolerance (D)
Black	0	0	1	1%
Brown	1	1	10	2%
Red	2	2	100	3%
Orange	3	3	1,000	4%
Yellow	4	4	10,000	
Green	5	5	100,000	
Blue	6	6	1,000,000	
Violet	7	7	10,000,000	
Gray	8	8	100,000,000	
White	9	9	10^9	
Gold			0.1 (EIA)	5%
Silver			0.01 (EIA)	10%
No Color				20%

Example: Red – Red – Orange = 22,000 ohms, 20%

Additional information concerning the Axial Lead resistor can be obtained if Band A is a wide band. Case 1: If only Bank A is wide, it indicates that the resistor is wirewound. Case 2: If Bank A is wide and there is also a blue fifth band to the right of Band D on the Axial Lead Resistor, it indicates the resistor is wirewound and flame proof.

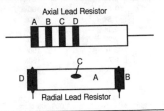

Axial Lead Resistor

Radial Lead Resistor

RESISTOR STANDARD VALUES

Standard Resistor Values for 5% class

1	62	3.9k	240k
1.1	68	4.3k	270k
1.2	75	4.7k	300k
1.3	82	5.1k	330k
1.5	91	5.6k	360k
1.6	100	6.2k	390k
1.8	110	6.8k	430k
2.0	120	7.5k	470k
2.2	130	8.2k	510k
2.4	150	9.1k	560k
2.7	160	10k	620k
3.0	180	11k	680k
3.3	200	12k	750k
3.6	220	13k	820k
3.9	240	15k	910k
4.3	270	16k	1.0M
4.7	300	18k	1.1M
5.1	330	20k	1.2M
5.6	360	22k	1.3M
6.2	390	24k	1.5M
6.8	430	27k	1.6M
7.5	470	30k	1.8M
8.2	510	33k	2.0M
9.1	560	36k	2.0M
10	620	39k	2.4M
11	680	43k	2.7M
12	750	47k	3.0M
13	820	51k	3.3M
15	910	56k	3.6M
16	1.0k	62k	3.9M
18	1.1k	68k	4.3M
20	1.2k	75k	4.7M
22	1.3k	82k	5.1M
24	1.5k	91k	5.6M
28	1.6k	100k	6.2M
30	1.8k	110k	6.8M
33	2.0k	120k	7.5M
36	2.2k	130k	8.2M
39	2.4k	150k	9.1M
43	2.7k	160k	10.0M
47	3.0k	180k	
51	3.3k	200k	
56	3.6k	220k	

k = kilohms = 1,000 ohms   M = megohms = 1,000,000 ohms

CAPACITOR COLOR CODES

Color	1st Digit (A)	2nd Digit (B)	Multiplier (C)	Tolerance (D)
Black	0	0	1	20%
Brown	1	1	10	1%
Red	2	2	100	2%
Orange	3	3	1,000	3%
Yellow	4	4	10,000	4%
Green	5	5	100,000	5%
Blue	6	6	1,000,000	6%
Violet	7	7	10,000,000	7%
Gray	8	8	100,000,000	8%
White	9	9	10^9	9%
Gold			0.1 (EIA)	5%
Silver			0.01 (EIA)	10%
No Color				20%

COLOR CODES FOR CERAMIC CAPACITORS

Color	Decimal Multiplier (C)	Tolerance (D) Above 10pf	Below 10pf	Temp Coef ppm/°C (E)
Black	1	20	20	0
Brown	10	1		-30
Red	100	2		-80
Orange	1000			-150
Yellow				-220
Green		5	0.5	-330
Blue				-470
Violet				-750
Gray	0.01		0.25	30
White	0.1	10	1.0	500

Ceramic disc capacitors are usually labeled. If the number is <1 then the value is picofarads, if >1 the value is microfarads. The letter R is sometimes used as a decimal, eg, 4R7 is 4.7.

CAPACITOR STANDARD VALUES

pF	mF	mF	mF	mF
10	0.001	0.1	10	1000
12	0.0012			
13	0.0013			
15	0.0015	0.15	15	
18	0.0018			
20	0.002			
22	0.0022	0.22	22	2200
24				
27				
30				
33	0.0033	0.33	33	3300
36				
43				
47	0.0047	0.47	47	4700
51				
56				
62				
68	0.0068	0.68	68	6800
75				
82				
100	0.01	1.0	100	10,000
110				
120				
130				
150	0.015	1.5		
180				
200				
220	0.022	2.2	220	22,000
240				
270				
300				
330	0.033	3.3	330	
360				
390				
430				
470	0.047	4.7	470	47,000
510				
560				
620				
680	0.068	6.8		
750				
820				82,000
910				

pF = picofarads = 1×10^{-12} farads   mF = microfarads = 1×10^{-6} farads

STANDARD WIRING CODES

Typically, the following color codes are used for **electronic applications** (as established by the Electronic Industries Association – EIA):

Wire Color (solid)	Circuit type
Black	Chassis grounds, returns, primary leads
Blue	Plate leads, transistor collectors, FET drain
Brown	Filaments, plate start lead
Gray	AC main power leads
Green	Transistor base, finish grid, diodes, FET gate
Orange	Transistor base 2, screen grid
Red	B plus dc power supply
Violet	Power supply minus
White	B – C minus of bias supply, AVC – AGC return
Yellow	Emitters-cathode and transistor, FET source

Stereo Audio Channels are color coded as follows:

Wire Color (solid)	Circuit type
White	Left channel high side
Blue	Left channel low side
Red	Right channel high side
Green	Right channel low side

AF Transformers (audio) are color coded as follows:

Wire Color (solid)	Circuit type
Black	Ground line
Blue	Plate, collector, or drain lead. End of primary winding.
Brown	Start primary loop. Opposite to blue lead.
Green	High side, end secondary loop.
Red	B plus, center tap push - pull loop.
Yellow	Secondary center tap.

IF Transformers (Intermediate Frequency) are color coded as follows:

Wire Color (solid)	Circuit type
Blue	Primary high side of plate, collector, or drain lead.
Green	Secondary high side for output.
Red	Low side of primary returning B plus.
Violet	Secondary outputs.
White	Secondary low side.

FUSES – SMALL TUBE TYPE

TYPE	Description	Diameter Inches	Length Inches
3AB	Ceramic body, normal, 200% 15sec	1/4	1-1/4
1AG	Auto Glass, fast blow, 200% 5sec	1/4	5/8
2AG	Auto Glass, fast blow, 200% 10sec	0.177	0.57
3AG	Auto Glass, fast blow, 200% 5sec	1/4	1-1/4
4AG	Auto Glass, fast blow, 200% 5sec	9/32	1-1/4
5AG	Auto Glass, fast blow, 200% 5sec	13/32	1-1/2
7AG	Auto Glass, fast blow, 200% 5sec	1/4	7/8
8AG	Auto Glass, fast blow, 200% 5sec	1/4	1
9AG	Auto Glass, fast blow, 200% 5sec	1/4	1-7/16
216	Metric, fast blow, high int., 210% 30m	5mm	20mm
217	Glass, Metric, fast blow, 210% 30m	5mm	20mm
218	Glass, Metric, slow blow, 210% 2 min	5mm	20mm
ABC	No Delay, Ceramic, 110% rating, Will blow at 135% load in one hour	1/4	1-1/4
AGC	Fast Acting, glass tube, 110% rating, Will blow at 135% load in one hour	1/4	1-1/4
AGX	Fast Acting, glass tube	1/4	1
BLF	No delay, 200% 15sec	13/32	1-1/2
BLN	No delay, military, 200% 15sec	13/32	1-1/2
BLS	Fast clearing, 600V, 135% 1hr	13/32	1-3/8
FLA	Time delay, indicator pin, 135% 1hr	13/32	1-1/2
FLM	Dual element, delay, 200% 12 sec	13/32	1-1/2
FLQ	Dual element, delay, 500V, 200% 12sec	13/32	1-1/2
FNM	Slow Blow Time Delay	13/32	1-1/4
FNA	Slow Blow, Indicator, silver pin pops out when blown, Dual Element	13/32	1-1/2
GBB	Rectifier Fuse, Fast, low let through	1/4	1-1/4
GLD	Indicator Fuse, Silver pin pops out to show blown fuse. 110% rating	1/4	1-1/4
GGS	Metric, fast acting	5mm	20mm
KLK	Fast, current limiting, 600V, 135% 1hr	13/32	1-1/2
KLW	Fast, protect solid state, 250% 1sec	13/32	1-1/2
MDL	Dual Element, Time Delay, glass tube	1/4	1-1/4
MDX	Dual Element, glass tube	1/4	1-1/4
MDV	Dual Element, glass tube, Pigtail	1/4	1-1/4
SC	Slow Blow, Time Delay Size rejection also	13/32	1-5/6 to 2-1/4
218000	Slow blow, glass body, 200% 5sec	0.197	0.787
251000	Pico II™ Subminiature, fast blow	Wire lead	
273000	Microfuse, fast blow, 200% 5sec	Wire lead	
313000	Slow blow, glass body, 200% 5sec	1/4	1-1/4
326000	Slow Blow, ceramic, 200% 5sec	1/4	1-1/4

Note: The 200% 10sec figures above indicate that a 200% overload will blow the fuse in 10 seconds.

BATTERY CHARACTERISTICS

Battery (1)	Anode	Cathode	Voltage (2)	Amp-hrs/kg
Ammonia	Mg	m-DNB	2.2 (1.7)	1,400
Cadmium-Air (C)	Cd	O_2	1.2 (0.8)	475
Cuprous chloride	Mg	CuCl	1.5 (1.4)	240
Edison (C)	Fe	NiO	1.5 (1.2)	195
H_2-O_2 (C)	H_2	O_2	1.23 (0.8)	3,000
Lead-Acid (C)	Pb	PbO_2	2.1 (2.0)	55
Leclanche (NC)	Zn	MnO_2	1.6 (1.2)	230
Lithium-High Temp, 350°C, with fused salt	Li	S	2.1 (1.8)	685
Magnesium (NC)	Mg	MnO_2	2.0 (1.5)	270
Mercury (NC)	Zn	HgO	1.34 (1.2)	185
Mercad (NC)	Cd	HgO	0.9 (0.85)	165
MnO_2 alkaline (NC)	Zn	MnO_2	1.5 (1.15)	230
NiCad (C)	Cd	NiO	1.35 (1.2)	165
Organic Cath. (NC)	Mg	m-DNB	1.8 (1.15)	1,400
Silver Cadmium (C)	Cd	AgO	1.4 (1.05)	230
Silver Chloride	Mg	AgCl	1.6 (1.5)	170
Silver Oxide	Zn	AgO	1.85 (1.5)	285
Silver-Poly	Ag	Polyiodide	0.66 (0.6)	180?
Sodium - High Temp, 300°C, with β-alumina electrolyte	Na	S	2.2 (1.8)	1,150
Thermal	Ca	Fuel	2.8 (2.6)	240
Zinc-Air (NC)	Zn	O_2	1.6 (1.1)	815
Zinc-Nickel (NC)	Zn	Ni oxides	1.75 (1.6)	185
Zinc-Silver Ox	Zn	AgO	1.85 (1.5)	285

Fuel Cells:

Battery (1)	Anode	Cathode	Voltage (2)	Amp-hrs/kg
Hydrogen	H_2	O_2	1.23 (0.7)	26,000
Hydrazine	N_2H_2	O_2	1.5 (0.7)	2,100
Methanol	CH_2OH	O_2	1.3 (0.9)	1,400

(1) (NC) after the name indicates the cell is a Primary Cell and cannot be recharged. (C) indicates the cell is a Secondary Cell and can be recharged.

(2) The first voltage is the theoretical voltage developed by the cell and the value in parenthesis is the typical voltage generated by a working cell. Amp-hrs/kg is the theoretical capacity of the cell.

Battery data listed above was obtained from the *Electronic Engineers Master Catalog, Hearst Business Communications Inc., 1986-1987.*

BATTERIES - STANDARD SIZES

Size	Eveready #	NEDA #	Voltage	Capacity
Carbon Zinc Cells:				
AAA	912	24F	1.5	20 ma @ 21 hrs
AA	915	15F	1.5	54 ma @ 20 hrs
C	935	14F	1.5	20 ma @ 140 hrs
C	1235	14D	1.5	37.5 ma @ 97 hrs
D	950	13F	1.5	20 ma @ 360 hrs
D	1150	13C	1.5	375 ma @ 15.8 hrs
D	1250	13D	1.5	60 ma @ 139 hrs
N	904	910F	1.5	20 ma @ 22 hrs
WO	201		1.5	0.1 ma @ 650 hrs
	750	704	3.0	20 ma @ 37 hrs
	715	903	4.5	120 ma @ 90 hrs
	724	2	6.0	60 ma @ 175 hrs
	509	908	6.0	187 ma @ 40 hrs
109	206	1611	9	12 ma @ 40 hrs
127	226	1600	9	12 ma @ 61 hrs
	276	1603	9	20 ma @ 350 hrs
117	216	1604	9	9 ma @ 50 hrs
	228	1810	12	12 ma @ 59 hrs
	420	225	22.5	5 ma @ 60 hrs
	482	207	45	40 ma @ 125 hrs
	490	204	90	10 ma @ 63 hrs
Alkaline – Manganese:				
AAA	E92	24A	1.5	37.5 ma @ 25 hrs
AA	E91	15A	1.5	20 ma @ 107 hrs
C	E93	14A	1.5	37.5 ma @ 160 hrs
D	E95	13A	1.5	50 ma @ 270 hrs
G	520	930A	6.0	375 ma @ 59 hrs
N	E90	910A	1.5	9 ma @ 90 hrs
	532	1308AP	3.0	20 ma @ 35 hrs
	531	1307AP	4.5	20 ma @ 35 hrs
	522	1604A	9.0	18 ma @ 33 hrs
	539		6.0	18 ma @ 30.5 hrs
Ni Cad Rechargeable:				
AAA	CH12ABP-2	10024	1.2	180 milliamp-hours
AA	CH15	10015	1.2	500 milliamp-hours
C	CH35	10014	1.2	1.2 ampere-hours
Sub C	CH1.2	10022	1.2	1.2 ampere-hours
D	CH50	10013	1.2	1.2 ampere-hours
D	CH4	10013HC	1.2	4 ampere-hours
N	CH150	10910	1.2	150 milliamp-hours
	CH22		8.4	80 milliamp-hours

CF series rechargeable is Fast Charge, CH is Standard Charge ma = milliamp
Note: CH should be charged at a rate of 1/10 amp-hour for 10 hours.

SINE WAVES

Frequency	Period	Peak-to-Peak Value
$f = \frac{1}{T}$	$T = \frac{1}{f}$	$V_{p\text{-}p} = 2 \times V_{max}$
where	where	where
f = frequency (in hertz)	T = period (in seconds)	2 = constant
1 = constant	1 = constant	$V_{p\text{-}p}$ = *peak-topeak value*
T = period (in seconds)	f = frequency (in hertz)	V_{max} = *peak value*

Average Value	rms Value
$V_{avg} = V_{max} \times .637$	$V_{rms} = V_{max} \times .707$
where	where
V_{avg} = average value (in volts)	V_{rms} = rms value (in volts)
V_{max} = *peak value* (in volts)	V_{max} = *peak value* (in volts)
.637 = constant	*.707* = constant

CONDUCTIVE LEAKAGE CURRENT

$$I_L = \frac{V_a}{R_1}$$

where

I_L = leakage current (in microamperes)

V_a = applied voltage (in volts)

R_1 = insulation resistance (in megohms)

PROGRAMMABLE CONTROLLER ERROR CODES

Error Code	Problem	Possible Cause	Corrective Action
01	Memory error occurring in run mode	Voltage surge, improper grounding, high noise interference on lines, inadequate power supply	Add surge suppression to incoming power lines and all inductive output lines. Check power supply voltage for correct level. Reload program with backup program and reboot system.
02	Processor does not meet required software level	Software is not compatible with the hardware system. Normally occurs when updated software is loaded into an older system	Consult the software specifications or manufacturer to determine the required hardware level. Upgrade system hardware to meet the software requirements
03	Power failure of an expansion I/O module	Power was removed from module, inadequate power supply, or power has dipped below the minimum specification of the module	Measure voltage at the module power supply and correct any problems. Reboot the system to return to normal operation when message still appears after power is restored
04	Program is trying to address an I/O module in an empty slot on rack	Program is set to the wrong rack and/or slot number	Set the program to a new address number or insert the correct I/O module in the slot number being addressed
05	Module has been detected as being inserted under power	Power was not disabled before a module was inserted into a slot	Reboot the system to return to normal operation when message still appears after power is restored. Never insert a module while power is applied to the unit. Always turn power OFF before inserting or removing a module

CAPACITOR RATINGS

110-125 VAC, 50/60 Hz, Starting Capacitors

Typical Ratings*	Dimensions**		Model Number*
	Diameter	Length	
88-106	1-7/16	2-3/4	EC8815
108-130	1-7/16	2-3/4	EC10815
130-489	1-7/16	2-3/4	EC13015
145-174	1-7/16	2-3/4	EC14815
161-193	1-7/16	2-3/4	EC16115
189-227	1-7/16	2-3/4	EC18915A
216-259	1-7/16	3-3/8	EC21615
233-280	1-7/16	3-3/8	EC23315A
243-292	1-7/16	3-3/8	EC24315A
270-324	1-7/16	3-3/8	EC27015A
324-389	1-7/16	3-3/8	EC2R10324N
340-408	1-13/16	3-3/8	EC34015
376-454	1-13/16	3-3/8	EC37815
400-480	1-13/16	3-3/8	EC40015
430-516	1-13/16	3-3/8	EC43015A
460-553	1-13/16	4-3/8	EC5R10460N
540-648	1-13/16	4-3/8	EC54015B
590-708	1-13/16	4-3/8	EC59015A
708-850	1-13/16	4-3/8	EC70815
815-978	1-13/16	4-3/8	EC81515
1000-1200	2-1/16	4-3/8	EC100015A

220-250 VAC, 50/60 Hz, Starting Capacitors

53-64	1-7/16	3-3/8	EC5335
64-77	1-7/16	3-3/8	EC6435
88-406	1-13/16	3-3/8	EC8835
108-130	1-13/16	3-3/8	EC10835A
124-149	1-13/16	4-3/8	EC12435
130-154	1-13/16	4-3/8	EC13035
145-174	2-1/16	3-3/8	EC6R22145N
161-193	2-1/16	3-3/8	EC6R2216N
216-259	2-1/16	4-3/8	EC21635A
233-280	2-1/16	4-3/8	EC23335A
270-324	2-1/16	4-3/8	EC27035A

* in μF
** in inches
*** Model numbers vary by manufacturer

CAPACITOR RATINGS

270 VAC, 50/60 Hz, Running Capacitors

Typical Ratings*	Dimensions**		Model Number*
	Oval	Length	
2		2-1/8	VH550
3		2-1/8	VH25503
4	1-5/16 x 2-5/32	2-1/8	VH5704
5		2-1/8	VH5705
6		2-5/8	VH5706
7.5		2-7/8	VH9001
10	1-5/16 x 2-5/32	2-7/8	VH9002
12.5		3-7/8	VH9003
15	1-29/32 x 2-29/32	3-1/8	VH9121
17.5		2-7/8	VH9123
20		2-7/8	VH5463
25	1-29/32 x 2-29/32	3-7/8	VH9069
30		3-7/8	VH5465
35	1-29/32 x 2-29/32	3-7/8	VH9071
40		3-7/8	VH9073
45	1-31/32 x 3-21/32	3-7/8	VH9115
50		3-7/8	VH9075

40 VAC, 50/60 Hz, Running Capacitors

Typical Ratings*	Dimensions**		Model Number*
	Oval	Length	
10	1-5/16 x 2-5/32	3-7/8	VH5300
15	1-29/32 x 2-29/32	2-7/8	VH5304
17.5	1-29/32 x 2-29/32	3-7/8	VH9141
20	1-29/32 x 2-29/32	3-7/8	VH9082
25	1-29/32 x 2-29/32	3-7/8	VH5310
30		4-3/4	VH9086
35	1-29/32 x 2-29/32	4-3/4	VH9088
40		4-3/4	VH9641
45		3-7/8	VH5351
50	1-31/32 x 3-21/32	3-7/8	VH5320
55		4-3/4	VH9081

* in μF
** in inches
*** Model numbers vary by manufacturer

RESISTOR COLOR CODES

Color	Number 1st	Number 2nd	Multiplier (C)	Tolerance (D)
Black (BK)	0	0	1	0
Brown (BR)	1	1	10	–
Red (R)	2	2	100	–
Orange (O)	3	3	1,000	–
Yellow (Y)	4	4	10,000	–
Green (G)	5	5	100,000	–
Blue (BL)	6	6	1,000,000	–
Violet (V)	7	7	10,000,000	–
Gray (GY)	8	8	100,000,000	–
White (W)	9	9	1,000,000,000	–
Gold (Au)	–	–	0.1	5
Silver (Ag)	–	–	0.01	10
None	–	–	0	20

ELECTROMAGNETIC FREQUENCY SPECTRUM

Frequency (Wavelength)	Name
0 Hertz	Steady direct current
15-20,000 Hz	Audio frequencies
30-15,000 Hz	Normal human hearing range
16-4186.01 Hz	Standard musical scales
Note: – Audio is mechanical - not electromagnetic	
3-30 Hz (100Mm)	e.l.f. – Extremely Low Frequency
30-300 Hz (10-1Mm)	u.l.f. – Ultra Low Frequency
3-30 kHz (100,000-10,000m)	v.l.f. – Very Low Frequency
10-16 kHz	ultrasonic
30 kHz to 30,000 MHz	Radio Frequencies
30-300 kHz (10,000-1000m)	l.f. – low frequencies
30-535 kHz	Marine com & navigation, aero nav.
300-3000 kHz (1000-100m)	m.f. – medium frequencies
535-1705 kHz	AM broadcast bands
1800-2000 kHz	Amateur band, 160 meter
3-30 MHz (100-10m)	h.f. – high frequencies
3.5-4 MHz	Amateur band, 80 meter
7-7.3 MHz	Amateur band, 40 meter
10.100-10.150 MHz	Amateur band, 30 meter
14.1-14.35 MHz	Amateur band, 20 meter
18.068-18.168 MHz	Amateur band, 17 meter
21-21.45 MHz	Amateur band, 15 meter
26.95-27.54 MHz	Industrial, scientific, & medical
24.890-24.990	Amateur band, 12 meter
28-29.7 MHz	Amateur band, 10 meter
26.965-27.405 MHz	Citizens Band (Class D)

ELECTROMAGNETIC FREQUENCY SPECTRUM

Frequency (Wavelength)	Name
30-300 MHz (10-1m)	v.h.f. – very high frequencies
30-50 MHz	Police, fire, forest, highway, railroad
50-54 MHz	Amateur band, 6 meter
54-72 MHz	TV channels 2 to 4
72-76 MHz	Government, Aero. Marker 75 MHz
76-88 MHz	TV channels 5 and 6
88-108 MHz	FM broadcast band
108-118 MHz	Aeronautical navigation
118-136 MHz	Civil Communication Band
148-174 MHz	Government
144-148 MHz	Amateur band, 2 meter
174-216 MHz	TV channels 7 to 13
216-470 MHz	Amateur, government, CB Band,
	• non-government, fixed or mobile
	• aeronautical navigation
220-225 MHz	Amateur band, 1-1/4 meter
225-400 MHz	Military
420-400 MHz	Amateur band, 0.7 meter
462.55-563.20 MHz	Citizens Band (Class A)
300-3000 MHz (100-10cm)	u.h.f. – ultra high frequencies
470-806 MHz	TV channels 14 to 69
806-890 MHz	Cellular telephone
890-3000 MHz	Aero navigation, amateur bands,
	• government & non-government
	• fixed and mobile
1300-1600	Radar band
3000-30,000 Mhz (10-1cm)	s.h.f. – super high frequencies
	• Government and non-government,
	• amateur bands, radio navigation
3000-30,000 MHz to 300 GHz (1-0.1cm)	Extra-high frequencies (weather radar, experimental, government)
30-0.76 μm	Infrared light and heat
0.76-0.39 μm	Visible light
6470-7000 ångstroms	Red light
5850-6740 ångstroms	Orange light
5750-5850 ångstroms	Yellow light
5560-5750 ångstroms	Maximum visibility
4912-5560 ångstroms	Green light
4240-4912 ångstroms	Blue light
4000-4240 ångstroms	Violet light
0.39-0.032 μm	Ultraviolet light
0.032-0.00001 μm	X-rays
0.00001-0.0000006 μm	Gamma rays
0.0005 ångstroms	Cosmic rays

LEAD–ACID BATTERY SPECIFIC GRAVITY & CHARGE

Acids Specific Gravity Charge Level

1.30 to 1.32	Overcharged
1.26 to 1.28	100%
1.24 to 1.26	75%
1.20 to 1.22	50%
1.15 to 1.17	25%
1.13 to 1.15	Very low capacity
1.11 to 1.12	Discharged

Battery Efficiency Changes with Temperature

80°F = 100% Charge	10°F = 50% Charge
50°F = 82% Charge	0°F = 40% Charge
30°F = 64% Charge	-10°F = 33% Charge
20°F = 58% Charge	-20°F = 18% Charge

MAGNETIC PERMEABILITY OF SOME COMMON MATERIALS

Substance	Permeability (approx.)
Aluminum	Slightly more than 1
Bismuth	Slightly less than 1
Cobalt	60-70
Ferrite	100-300
Free space	1
Iron	60-100
Iron, refined	3000-8000
Nickel	50-60
Permalloy	3000-30,000
Silver	Slightly less than 1
Steel	300-600
Super permalloys	100,000-1,000,000
Wax	Slightly less than 1
Wood, dry	Slightly less than 1

TRANSISTOR CIRCUIT ABBREVIATIONS

Quantity	Abbreviations
Base-emitter voltage	E_B, V_B, E_{BE}, V_{BE}
Collector-emitter voltage	E_C, V_C, E_{CE}, V_{CE}
Collector-base voltage	E_{BC}, V_{BC}, E_{CB}, V_{CB}
Gate-source voltage	E_G, V_G, E_{GS}, V_{GS}
Drain-source voltage	E_D, V_D, E_{DS}, V_{DS}
Drain-gate voltage	E_{DG}, V_{DG}, E_{GD}, V_{GD}
Emitter current	I_E
Base current	I_B, I_{BE}, I_{EB}
Collector current	I_C, I_{CE}, I_{EC}
Source current	I_S
Gate current	I_G, I_{GS}, I_{SG}*
Drain current	I_D, I_{DS}, I_{SD}

*This is almost always insignificant.

RADIO FREQUENCY CLASSIFICATIONS

Classification	Abbreviation	Frequency range
Very Low Frequency	VLF	9 kHz and below
Low Frequency (Longwave)	LF	30 kHz - 300 kHz
Medium Frequency	MF	300 kHz - 2 MHz
High Frequency (Shortwave)	HF	3 MHz - 30 MHz
Very High Frequency	VHF	30 MHz - 300 MHz
Ultra High Frequency	UHF	300 MHz - 3 GHz
Microwaves		3 GHz and above

ELECTRONIC SYMBOLS

ammeter	
amplifier (operational)	
AND gate	
antenna (balanced, dipole)	
antenna (general)	
antenna (loop, shielded)	
antenna (loop, unshielded)	
antenna (unbalanced)	
antenna (whip)	
attenuator (or resistor, fixed)	
attenuator (or resistor, variable)	
battery	
capacitor (feedthrough)	
capacitor (fixed, nonpolarized)	
capacitor (fixed, polarized)	
capacitor (ganged, variable)	
capacitor (single variable)	
capacitor (split-rotor, variable)	
capacitor (split-stator, variable)	
cathode (directly heated)	
cathode (indirectly heated)	
cathode (cold)	
cavity resonator	
cell	
circuit breaker	

	coaxial cable
	coaxial cable (grounded shield)
	crystal (piezoelectric)
	delay line
	diode (field effect)
	diode (general)
	diode (Gunn)
	diode (light-emitting)
	diode (photosensitive)
	diode (photovoltaic)
	diode (pin)
	diode (Schottky)
	diode (tunnel)
	diode (varactor)
	diode (zener)
	directional coupler (or wattmeter)
	exclusive-OR gate
	female contact (general)
	ferrite bead
	fuse
	galvanometer
	ground (chassis)
	ground (earth)
	handset

headphone (single)

headphone (stereo)

inductor (air-core)
inductor (bifilar)
inductor (iron-core)
inductor (tapped)

inductor (variable)

integrated circuit

inverter or inverting amplifier

jack (coaxial or phono)

jack (phone, two-conductor)

jack (phone, two-conductor interrupting)

jack (phone, three-conductor)

jack (phono)

key (telegraph)

lamp (incandescent)

lamp (neon)

male contact (general)

meter (general)

microammeter

microphone

microphone (directional)

milliammeter

NAND gate

negative voltage connection

NOR gate

operational amplifier

OR gate

outlet (nonpolarized)

outlet (polarized)

outlet (utility, 117 V, nonpolarized)

outlet (utility, 234 V)

photocell (tube)

plug (nonpolarized)

plug (polarized)

plug (phone, two-conductor)

plug (phone, three-conductor)

plug (phono)

plug (utility, 117 V)

plug (utility, 234 V)

positive-voltage connection

potentiometer (variable resistor, or rheostat)

probe (radio-frequency)

rectifier (semiconductor)

rectifier (silicon-controlled)

rectifier (tube-type)

ELECTRONIC SYMBOLS (CONT.)

rectifier (tube-type, gas-filled)

relay (DPDT)

relay (DPST)

relay (SPDT)

relay (SPST)

resistor (fixed)

resistor (preset)

resistor (tapped)

resonator

rheostat (variable resistor, or potentiometer)

saturable reactor

shielding

signal generator

solar cell source (constant-voltage)

source (constant-current)

speaker

switch (DPDT)

switch (DPST)

switch (momentary-contact)

ELECTRONIC SYMBOLS (CONT.)

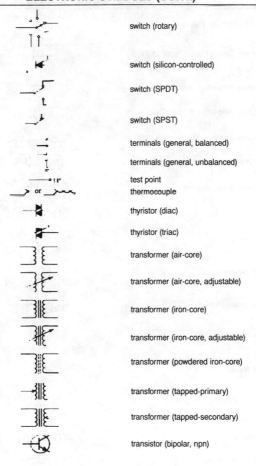

switch (rotary)

switch (silicon-controlled)

switch (SPDT)

switch (SPST)

terminals (general, balanced)

terminals (general, unbalanced)

test point

thermocouple

thyristor (diac)

thyristor (triac)

transformer (air-core)

transformer (air-core, adjustable)

transformer (iron-core)

transformer (iron-core, adjustable)

transformer (powdered iron-core)

transformer (tapped-primary)

transformer (tapped-secondary)

transistor (bipolar, npn)

transistor (bipolar, pnp)

transistor (junction field-effector, JFET)

transistor (field-effect, n-channel)

transistor (field-effect, P-channel)

transistor (metal-oxide, dual-gate)

transistor (metal-oxide, single-gate)

transistor (photosensitive)

transistor (unijunction)

tube (diode)

tube (pentode)

tube (photomultiplier)

tube (tetrode)

tube (triode)

unspecified unit or component

voltmeter

wattmeter

waveguide (circular)

ELECTRONIC SYMBOLS (CONT.)

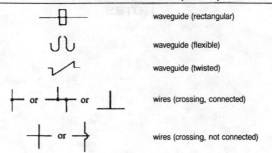

waveguide (rectangular)

waveguide (flexible)

waveguide (twisted)

wires (crossing, connected)

wires (crossing, not connected)

NOTES

CHAPTER 3
RACEWAYS & WIRING

TYPICAL POWER WIRING COLOR CODE

120/240 Volt		277/480 Volt	
Black	Phase 1	Brown	Phase 1
Red	Phase 2	Orange	Phase 2
Blue	Phase 3	Yellow	Phase 3
White	Neutral	Gray	Neutral
Green	Ground	Green w/yellow stripe	Ground

POWER-TRANSFORMERS ARE COLOR CODED AS FOLLOWS:

Wire Color (solid)	Circuit Type
Black	If a transformer does not have a tapped primary, both leads are black.
Black	If a transformer does have a tapped primary, the black is the common lead.
Black & Yellow.	Tap for a tapped primary.
Black & Red. . .	End for a tapped primary.

CONDUIT SIZE vs WIRE SIZE

Wire Size AWG	Minimum Conduit Size (inches) per Number of Type TW Wires. Number of Wires Inside Conduit				
	2	3	4	5	6
14	1/2	1/2	1/2	1/2	1/2
12	1/2	1/2	1/2	1/2	1/2
10	1/2	1/2	1/2	1/2	3/4
8	1/2	3/4	3/4	1	1
6	3/4	1	1	1-1/4	1-1/4
4	1	1	1-1/4	1-1/4	1-1/2
2	1	1-1/4	1-1/4	1-1/2	2
1/0	1-1/4	1-1/2	2	2	2-1/2
2/0	1-1/2	1-1/2	2	2	2-1/2
3/0	1-1/2	2	2	2-1/2	2-1/2

See the National Electric Code for conduit sizes when using wire types other than TW.

BOX SIZE vs NUMBER OF WIRES

Maximum Number of Wires in a Junction Box
Wire Size AWG

Box Size in inches	#14	#12	#10	#8
Outlet Boxes				
4-11/16 x 1-1/4 square	12	11	10	8
4-11/16 x 1-1/2 square	14	13	11	9
4-11/16 x 2-1/8 square	21	18	16	14
4 x 1-1/4 octagon or round	6	5	5	4
4 x 1-1/2 octagon or round	7	6	6	5
4 x 2-1/8 octagon or round	10	9	8	7
4 x 1-1/4 square	9	8	7	6
4 x 4 x 1-1/2	10	9	8	7
4 x 4 x 2-1/8	15	13	12	10
Switch Boxes				
3 x 2 x 1-1/2	3	3	3	2
3 x 2 x 2	5	4	4	3
3 x 2 x 2-1/4	5	4	4	3
3 x 2 x 2-1/2	6	5	5	4
3 x 2 x 2-3/4	7	6	5	4
3 x 2 x 3-1/2	9	8	7	6
4 x 2-1/8 x 1-1/2	5	4	4	3
4 x 2-1/8 x 1-7/8	6	5	5	4
4 x 2-1/8 x 2-1/8	7	6	5	4

The above numbers are maximums and you should deduct 1 wire for each outlet, switch, cable clamp, fixture-stud, or similar part that is also installed in the box.

RESISTIVITIES OF DIFFERENT SOILS

Soil	Resistivity OHM-CM		
	Average	Mn.	Max.
Fills – ashes, cinders, brine wastes	2,370	590	7,000
Clay, shale, gumbo, loam	4,060	340	16,300
Same – with varying proportions of sand and gravel	15,800	1,020	135,000
Gravel, sand, stones, with little clay or loam	94,000	59,000	458,000

TIGHTENING TORQUE IN POUND-FEET-SCREW FIT

Wire Size, AWG	Driver	Bolt	Other
18-16	1.67	6.25	4.2
14-8	1.67	6.25	6.125
6-4	3.0	12.5	8.0
3-1	3.2	21.00	10.40
0-2/0	4.22	29	12.5
AWG200MCM	–	37.5	17.0
250-300	–	50.0	21.0
400	–	62.5	21.0
500	–	62.5	25.0
600-750	–	75.0	25.0
800-1000	–	83.25	33.0
1250-2000	–	83.26	42.0

SCREWS

Screw Size, Inches Across, Hex Flats	Torque, Pound-Feet
1/8	4.2
5/32	8.3
3/16	15
7/32	23.25
1/4	42

WIRE SIZE vs VOLTAGE DROP

Voltage drop is the amount of voltage lost over the length of a piece of wire. Voltage drop changes as a function of the resistance of the wire and should be less than 2% if possible. If the drop is greater than 2%, efficiency of the appliance is severely decreased and life of the equipment will be decreased. As an example, if the voltage drop on an incandescent light bulb is 10%, the light output of the bulb decreases over 30%!

Voltage drop can be calculated using Ohm's Law, which is Voltage Drop = Current in amps x Resistance in ohms. For example, the voltage drop over a 200 foot long, 14 gauge power line supplying a 1000 watt floodlight is calculated as follows:

Current = 1000 watts / 120 volts = 8.4 amps
Resistance of #14 wire = 2.58 ohms / 1000 feet @ 77°F
Resistance of power line = 200 feet x 0.00258 ohms/foot
                        = 0.516 ohms
Voltage drop = 8.4 amps x 0.516 ohms = 4.33 volts
Percent voltage drop = 4.33 volts / 120 volts = 3.6%

The 3.6% drop is over the maximum 2% so either the wattage of the bulbs must be decreased or the diameter of the wire must be increased (a decrease in wire gauge number). If #12 wire were used in the above example, the voltage drop would have only been 2.2%. The wire resistance values for various size wire are contained in the Copper Wire table on page 114.

An interesting corollary to the above example is that if the line voltage doubles (240 volts instead of 120 volts) the voltage drop decreases by 50%. That means that a line can carry the same power 2 times further! Higher voltage lines are more efficient.

A more commonly used method of calculating voltage drop is as follows:

$$\text{Voltage drop} = \frac{22 \times \text{Wire length in feet} \times \text{current in amps}}{\text{Circular Mils}}$$

Using the values in the Ohm's Law example at the top of this page,
then Voltage drop = (22 x 200 x 8.4) / 4110 circ. mils = 9 volts = 7.5%.

Circular mils are given in the Standard Copper Wire Specs table on page 115. Note that the 22 value applies to copper wire only, if aluminum is used, change the value to 36.

WIRE SIZE vs VOLTAGE DROP

Max Wire Feet @ <u>120</u> Volts, 1 Phase, 2% Max Voltage Drop

Amps	Volt-Amps	#14	#12	#10	#8	#6
1	120	450	700	1100	1800	2800
5	600	90	140	225	360	575
10	1200	45	70	115	180	285
15	1800	30	47	75	120	190
20	2400	–	36	57	90	140
25	3000	–	–	45	72	115
30	3600	–	–	38	60	95
40	4800	–	–	–	45	72
50	6000	–	–	–	–	57

Amps	Volt-Amps	#4	#2	1/0	2/0	3/0
1	120	4500	7000	–	–	–
5	600	910	1400	2250	2800	–
10	1200	455	705	1100	1400	1800
15	1800	305	485	770	965	1200
20	2400	230	365	575	725	900
25	3000	180	290	460	580	720
30	3600	150	240	385	490	600
40	4800	115	175	290	360	440
50	6000	90	145	230	290	360
60	7200	76	120	190	240	305
70	8400	65	105	165	205	260
80	9600	–	90	144	180	230

Max Wire Feet @ <u>240</u> Volts, 1 Phase, 2% Max Voltage Drop

Amps	Volt-Amps	#14	#12	#10	#8	#6
1	240	900	1400	2200	3600	5600
5	1200	180	285	455	720	1020
10	2400	90	140	225	360	525
15	3600	60	95	150	240	350
20	4800	–	70	110	180	265
25	6000	–	–	90	144	210
30	7200	–	–	75	120	175
40	5600	–	–	–	90	130
50	12000	–	–	–	–	105

Amps	Volt-Amps	#4	#2	1/0	2/0	3/0
1	240	9000	–	–	–	-
5	1200	1750	2800	4500	5600	7000
10	2400	910	1400	2200	2800	3600
15	3600	605	965	1500	1900	2400
20	4800	455	725	1100	1400	1800
25	6000	365	580	920	1100	1440
30	7200	300	485	770	970	1200
40	5600	230	360	575	725	880
50	12000	180	290	460	580	720
60	14400	150	240	385	485	600
70	16800	130	205	330	415	520
80	19200	–	180	290	365	440
100	24000	–	–	230	280	360
150	36000	–	–	185	190	240
200	48000	–	–	–	–	180

WIRE CLASSES & INSULATION

Standard cable, as used in home and general construction, is classified by the wire size, number of wires, insulation type and dampness condition of the wire environment. Example: a cable with the code "12/2 with Ground – Type UF – 600V – (UL)" has the following specifications:

Wire size is 12 gauge (minimum required size for homes today; see the National Electric Code).

The "/2" indicates there are two wires in the cable.

"Ground" indicates there is a third wire in the cable to be used as a grounding wire.

"Type UF" indicates the insulation type and acceptable dampness rating.

"600V" means the wire is rated at 600 volts maximum.

"UL" indicates the wire has been certified by Underwriters Laboratory to be safe.

Cables are dampness rated as follows:

DRY: No dampness normally encountered. Indoor location above ground level.

DAMP: Partially protected locations. Moderate amount of moisture. Indoor location below ground level.

WET: Water saturation probable, such as underground or in concrete slabs or outside locations exposed to weather.

There are literally hundreds of different types of insulation used in wire and cable. To make things simple, the following descriptions are for wires commonly used in home wiring:

"BX" Armor covered with flexible, galvanized steel. Normally used in dry locations. Not legal to use in some states such as California.

"ROMEX" Although actually a trade name, it is used to describe a general class of plastic coated cable. Each wire is plastic wrapped except possibly the ground wire, which is sometimes bare or paper covered. Very flexible. There are three general types:

"NM" – Dry only, 2 or 3 wire, ground wire plastic wrapped.

"NMC" – Dry, 2 or 3 wire, all wires in solid plastic.

"UF" – Wet, 2 or 3 wire, all wires in solid, water resistant plastic. Use also instead of conduit.

WIRE CLASSES & INSULATION

Wire types are typically coded by the type of insulation, temperature range, dampness rating, and type and composition of the jacket. The following are some of the "Type Codes."

"T..."   Very common, dry only, full current load temperature must be less than 60°C (140°F).

"F"   Fixture wire. CF has cotton insulation (90°C), AF has asbestos insulation (150°C), SF has silicone insulation (200°C).

"R..."   Rubber (natural, neoprene, etc.) covered.

"S..."   Appliance cord, stranded conductors, cotton layer between wire and insulation, jute fillers, rubber outer jacket. S is extra hard service, SJ lighter service, SV light service.

"SP..."   Lamp cord, rubber insulation.

"SPT..."   Lamp cord, plastic insulation.

"X..."   Insulation is a cross linked synthetic polymer. Very tough and heat and moisture resistant.

"FEP..."   Fluorinated ethylene propylene insulation. Rated over 90°C (194°F). Dry only.

"...B"   Suffix indicating an outer braid is used, such as glass.

"...H"   Suffix indicating Higher loaded current temperatures may be used, up to 75°C (167°F).

"...HH"   Suffix indicating much higher loaded current temperatures may be used, up to 90°C (194°F).

"...L"   Suffix indicating a seamless lead jacket.

"...N"   Suffix indicating the jacket is extruded nylon or thermoplastic polyester and is very resistant to gas and oil and is very tough.

"...O"   Suffix indicating neoprene jacket.

"...W"   Suffix indicating WET use type.

Examples of some of the more common wire types are "T", "TW", "THWN", "THHN", "XHHW", "RHH", and "RHW".

COPPER WIRE RESISTANCE

Gauge A.W.G.	Feet per Ohm @ 77°F	Ohms per 1000 ft. @ 77°F	Feet per Ohm @ 149°F	Ohms per 1000 ft. @ 149°F
0000	20000	0.050	17544	0.057
000	15873	0.063	13699	0.073
00	12658	0.079	10870	0.092
0	10000	0.100	8621	0.116
1	7936	0.126	6849	0.146
2	6289	0.159	5435	0.184
3	4975	0.201	4310	0.232
4	3953	0.253	3425	0.292
5	3135	0.319	27.10	0.369
6	2481	0.403	2151	0.465
7	1968	0.508	1706	0.586
8	1560	0.641	1353	0.739
9	1238	0.808	1073	0.932
10	980.4	1.02	847.5	1.18
11	781.3	1.28	675.7	1.48
12	617.3	1.62	534.8	1.87
13	490.2	2.04	423.7	2.36
14	387.6	2.58	336.7	2.97
15	307.7	3.25	266.7	3.75
16	244.5	4.09	211.4	4.73
17	193.8	5.16	167.8	5.96
18	153.6	6.51	133.2	7.51
19	121.8	8.21	105.5	9.48
20	96.2	10.4	84.0	11.9
21	76.3	13.1	66.2	15.1
22	60.6	16.5	52.6	19.0
23	48.1	20.8	41.7	24.0
24	38.2	26.2	33.1	30.2
25	30.3	33.0	26.2	38.1
26	24.0	41.6	20.8	48.0
27	19.0	52.5	16.5	60.6
28	15.1	66.2	13.1	76.4
29	12.0	83.4	10.4	96.3
30	9.5	105	8.3	121
31	7.5	133	6.5	153
32	6.0	167	5.2	193
33	4.7	211	4.1	243
34	3.8	266	3.3	307
35	3.0	335	2.6	387
36	2.4	423	2.0	488
37	1.9	533	1.6	616
38	1.5	673	1.3	776
39	1.2	848	1.0	979
40	0.93	1070	0.81	1230

STANDARD COPPER WIRE SPECS

Gauge A.W.G.*	Diameter in mils (1000th in)	Diameter Millimeters	Area in Circular Mils	Weight Lbs. per 1000 feet	Turns /inch Enamel
0000	460.0	11.684	212000	641.0	2.2
000	410.0	10.414	168000	508.0	2.4
00	365.0	9.271	133000	403.0	2.7
0	325.0	8.255	106000	319.0	3.0
1	289.0	7.348	83700	253.0	3.3
2	258.0	6.544	66400	201.0	3.8
3	229.0	5.827	52600	159.0	4.2
4	204.0	5.189	41700	126.0	4.7
5	182.0	4.621	33100	100.0	5.2
6	162.0	4.115	26300	79.5	5.9
7	144.0	3.665	20800	63.0	6.5
8	128.0	3.264	16500	50.0	7.6
9	114.0	2.906	13100	39.6	8.6
10	102.0	2.588	10400	31.4	9.6
11	91.0	2.305	8230	24.9	10.7
12	81.0	2.053	6530	19.8	12.0
13	72.0	1.828	5180	15.7	13.5
14	64.0	1.628	4110	2.4	15.0
15	57.0	1.450	3260	9.86	16.8
16	51.0	1.291	2580	7.82	18.9
17	45.0	1.150	2050	6.2	21.2
18	40.0	1.024	1620	4.92	23.6
19	36.0	0.912	1290	3.90	26.4
20	32.0	0.812	1020	3.09	29.4
21	28.5	0.723	810	2.45	33.1
22	25.3	0.644	642	1.94	37.0
23	22.6	0.573	509	1.54	41.3
24	20.1	0.511	404	1.22	46.3
25	17.9	0.455	320	0.970	51.7
26	15.9	0.405	254	0.769	58.0
27	14.2	0.361	202	0.610	64.9
28	12.6	0.321	160	0.484	72.7
29	11.3	0.286	127	0.384	81.6
30	10.0	0.255	101	0.304	90.5
31	8.9	0.227	79.7	0.241	101
32	8.0	0.202	63.2	0.191	113
33	7.1	0.180	50.1	0.152	127
34	6.3	0.160	39.8	0.120	143
35	5.6	0.143	31.5	0.095	158
36	5.0	0.127	25.0	0.0757	175
37	4.5	0.113	19.8	0.0600	198
38	4.0	0.101	15.7	0.0476	224
39	3.5	0.090	12.5	0.0377	248
40	3.1	0.080	9.9	0.0200	282

*American Wire Gauge (formerly Brown & Sharp)

ALUMINUM WIRE AMP CAPACITY
Single wire in open air, ambient temp 86°F

Ampacities of Wire Types (w/Temp Rating) @ 0-2000 Volts

Wire Size AWG	UF TW (140°F)	RH, RHW THW, THWN XHHW, THHW (167°F)	THWN-2, XHH, USE-2 TA, TBS, SA, THHW, SIS, RHH, THW-2, THHN, XHHW, RHW-2 XHHW-2, ZW-2 (194°F)
500kcmil	405	485	545
400kcmil	355	425	480
300kcmil	290	350	395
0000	235	280	315
000	200	240	275
00	175	210	235
0	150	180	205
1	130	155	175
2	110	135	150
3	95	115	130
4	80	100	110
6	60	75	80
8	45	55	60
10	35	40	40
12	25	30	35

Note: Type TW is the most common for house wiring. TW is for dry or wet conditions and is covered with a single layer of plastic.
This table also applies to copper-clad aluminum
If the ambient [1] temperature is over 86°F (30°C), then the following corrections should be applied by multiplying the above ampacities by the correction factor below. kcmil = 1000 circular mil

Ambient [1] Temp °F	Ampacity Correction for above Wire Types		
	140°F	167°F	194°F
96-104	0.82	0.88	0.91
105-113	0.71	0.82	0.87
114-122	0.58	0.75	0.82
123-131	0.41	0.67	0.76
132-140	–	0.58	0.71
141-158	–	0.33	0.58
159-176	–	–	0.41

Note: The information on pages 112 and 113 has been extracted from the National Electrical Code ®, National Fire Protection Association, Quincy, Massachusetts 02269, Copyright 1993 and does not represent the complete code.

[1] Ambient temperature is the temperature of the material (air, earth, etc.) surrounding the wire.

ALUMINUM WIRE AMP CAPACITY
Three wires in cable, ambient temp 86°F

Ampacities of Wire Types (w/Temp Rating) @ 0-2000 Volts

Wire Size AWG	UF TW (140°F)	RH, RHW THW, THWN XHHW, THHW (167°F)	THWN-2, XHH, USE-2 TA, TBS, SA, THHW, SIS, RHH, THW-2, THHN, XHHW, RHW-2 XHHW-2, ZW-2 (194°F)
500kcmil	260	310	350
400kcmil	225	270	305
300kcmil	190	230	255
0000	150	180	205
000	130	155	175
00	115	135	150
0	100	120	135
1	85	100	115
2	75	90	100
3	65	75	85
4	55	65	75
6	40	50	60
8	30	40	45
10	25	30	35
12	20	20	25

Note: All notes on ambient temperature and TW types on the previous page also apply to this Three Wire section. kcmil = 1000 circular mil.
This table also applies to copper-clad aluminum

CURRENT ADJUSTMENT FOR MORE THAN 3 WIRES IN A CABLE

Number of Conductors	Percentage of amperage vlaue listed in amperage tables on the previous 4 pages
4 to 6	80%
7 to 9	70%
10 to 20	50%
21 to 30	45%
31 to 40	40%
over 41	35%

Basically, the above table reflects the rule that the higher the temperature (more wires = higher temperature) the lower the current carrying capacity of the wire.

Note: In all Aluminum and Copper Clad Aluminum Wire Types listed on pages 112 and 113 (except Types TA, TBS, SA, SIS, THW-2, THWN-2, RHW-2, USE-2, XHH, XHHW-2 & ZW-2) overcurrent protection should not exceed 15 amps for 12 AWG and 25 amps for 10AWG. This is not true if specifically permitted elsewhere in the Code.

COPPER WIRE CURRENT CAPACITY
Single wire in open air, ambient temp 86°F

Ampacities of Wire Types (w/Temp Rating) @ 0-2000 Volts

Wire Size AWG	TW UF (140°F)	FEPW, RH, RHW THW, THWN, ZW XHHW, THHW (167°F)	USE-2, XHH, XHHW-2 TA, TBS, SA, SIS, FEP, MI, RHW-2, THHN, ZW-2 THWN-2, FEPB, RHH THHW, THW-2 (194°F)
0000	300	360	405
000	260	310	350
00	225	265	300
0	195	230	260
1	165	195	220
2	140	170	190
3	120	145	165
4	105	125	140
6	80	95	105
8	60	70	80
10	40	50	55
12	30	35	40
14	25	30	35
16	–	–	24
18	–	–	18

If the ambient [1] temperature is over 86°F (30°C), then the following corrections should be applied by multiplying the above ampacities by the correction factor below.

Ambient [1] Temp °F	Ampacity Correction for above Wire Types		
	140°F	167°F	194°F
96-104	0.82	0.88	0.91
105-113	0.71	0.82	0.87
114-122	0.58	0.75	0.82
123-131	0.41	0.67	0.76
132-140	–	0.58	0.71
141-158	–	0.35	0.58
159-176	–	–	0.41

Note: The information on pages 110 and 111 has been extracted from the National Electrical Code ®, National Fire Protection Association, Quincy, Massachusetts 02269, Copyright 1993 and does not represent the complete code.

[1] Ambient temperature is the temperature of the material (air, earth, etc.) surrounding the wire.

COPPER WIRE CURRENT CAPACITY
Three wires in cable, ambient temp 86°F

Ampacities of Wire Types (w/Temp Rating) @ 0-2000 Volts

Wire Size AWG	TW UF (140°F)	FEPW, RH, RHW THW, THWN, ZW XHHW, THHW (167°F)	USE-2, XHH, XHHW-2 TA, TBS, SA, SIS, FEP, MI, RHW-2, THHN, ZW-2 THWN-2, FEPB, RHH THHW, THW-2 (194°F)
0000	195	230	260
000	165	200	225
00	145	175	195
0	125	150	170
1	110	130	150
2	95	115	130
3	85	100	110
4	70	85	95
6	55	65	75
8	40	50	65
10	30	35	40
12	25	25	30
14	20	20	25
16	–	–	18
18	–	–	14

STANDARD LAMP & EXTENSION CORD CURRENT CAPACITIES

Wire Size AWG	Wire Types SP, SPT, S, SJ, SV, ST, SJT, SVT		
	2 Conductor	3 Conductor	4 Conductor
10	30	25	20
12	25	20	16
14	18	15	12
16	13	10	8
18	10	7	6

Note: In all Copper Wire Types listed on pages 110 and 111 (except Types MI, TA, TBS, SA, SIS, RHW-2, THW-2, THWN-2, USE-2, XHH, XHHW-2, & ZW-2) overcurrent protection should not exceed 15 amps for 14 AWG, 20 amps for 12 AWG, and 30 amps for 10 AWG. This is not true if specifically permitted elsewhere in the Code.

ENCLOSURES

Type	Service Conditions	Sealing Method	Cost
1	No unusual		Base
4	Windblown dust and rain, splashing water, hose-directed water, and ice on enclosure		12 x Base
4X	Corrosion, windblown dust and rain, splashing water, hose-directed water, and ice on enclosure.		12 x Base
7	Withstand and contain an internal explosion of specified gases, contain an explosion sufficiently so an explosive gas-air mixture in the atmosphere is not ignited.		48 x Base
9	Dust		48 x Base
12	Dust, falling dirt, and dripping noncorrosive liquids		5 x Base

HAZARDOUS LOCATIONS

Class	Group	Material
I	A	Acetylene
	B	Hydrogen, butadiene, ethylene oxide, propylene oxide
	C	Carbon monoxide, ether, ethylene, hydrogen sulfide, morpholine, cyclopropane
	D	Gasoline, benzene, butane, propane, alcohol, acetone, ammonia, vinyl chloride
II	E	Metal dusts
	F	Carbon black, coke dust, coal
	G	Grain dust, flour, starch, sugar, plastics
III	No groups	Wood chips, cotton, flax, and nylon

ENCLOSURE TYPES

Type	Use	Service Conditions	UL Tests	Comments
1	Indoor	None	Rod entry, rust resistance	
3	Outdoor	Windblown dust, rain, sleet, and ice on enclosure	Rain, external icing, dust, and rust resistance	Do not provide protection against internal condensation, or internal icing
3R	Outdoor	Falling rain and ice on enclosure	Rod entry, rain, external icing, and rust resistance	Do not provide protection against dust, internal condensation, or internal icing
4	Indoor/outdoor	Windblown dust and rain, splashing water, hose-directed water and ice on enclosure	Hosedown, external icing and rust resistance	Do not provide protection against internal condensation, or internal icing
4X	Indoor/outdoor	Corrosion, windblown dust and rain, splashing water, hose-directed water and ice on enclosure	Hosedown, external icing, and corrosion resistance	Do not provide protection against internal condensation, or internal icing
6	Indoor/outdoor	Occasional temporary submersion at a limited depth		
6P	Indoor/outdoor	Prolonged submersion at a limited depth		

ENCLOSURE TYPES

Type	Use	Service Conditions	UL Tests	Comments
7	Indoor locations classified as Class I, or Groups A, B, C, or D, as defined in the NEC®	Withstand and contain an internal explosion of specified gases, contain an explosion sufficiently so an explosive gas-air mixture in the atmosphere is not ignited	Explosion, hydrostatic, and temperature	Enclosed heat-generating devices shall not cause external surfaces to reach temperatures capable of igniting explosive gas-air mixtures in the atmosphere
9	Indoor locations classified as Class II, Groups E or G, as defined in the NEC®	Dust	Dust penetration, temperature, and gasket aging	Enclosed heat-generating devices shall not cause external surfaces to reach temperatures capable of igniting explosive gas-air mixtures in the atmosphere
12	Indoor	Dust, falling dirt, and dripping noncorrosive liquids	Drip, dust, and rust resistance	Do not provide protection against internal condensation
13	Indoor	Dust, spraying water, oil and noncorrosive coolant	Oil explosion and rust resistance	Do not provide protection against internal condensation

POWER-FACTOR IMPROVEMENT
Capacitor Multipliers for Kilowatt Load

To give capacitor kvar required to improve power-factor from original to desired value - see sample below

Original Power Factor Percent	Desired Power Factor - Percent				
	100	95	90	85	80
60	1.333	1.004	0.849	0.713	0.583
62	1.266	0.937	0.782	0.646	0.516
64	1.201	0.872	0.717	0.581	0.451
66	1.138	0.809	0.654	0.518	0.388
68	1.078	0.749	0.594	0.458	0.338
70	1.020	0.691	0.536	0.400	0.270
72	0.964	0.635	0.480	0.344	0.214
74	0.909	0.580	0.425	0.289	0.159
76	0.855	0.526	0.371	0.235	0.105
77	0.829	0.500	0.345	0.209	0.079
78	0.802	0.473	0.318	0.182	0.052
79	0.776	0.447	0.292	0.156	0.026
80	0.750	0.421	0.266	0.130	
81	0.724	0.395	0.240	0.104	
82	0.698	0.369	0.214	0.078	
83	0.672	0.343	0.188	0.052	
84	0.646	0.317	0.162	0.206	
85	0.620	0.291	0.136		
86	0.593	0.264	0.109		
87	0.567	0.238	0.083		
88	0.540	0.211	0.056		
89	0.512	0.183	0.028		
90	0.484	0.155			
91	0.456	0.127			
92	0.426	0.097			
93	0.395	0.066			
94	0.363	0.034			
95	0.329				
96	0.292				
97	0.251				
99	0.143				

Assume the total plant load is 100kw at 60 percent power factor. Capacitor kvar rating necessary to improve power factor to 80 percent is found by multiplying kw (100) by multiplier in table (0.583), which gives kvar (58.3). Nearest standard rating (60 kvar) should be recommended.

MAXIMUM NUMBER OF CONDUCTORS IN CONDUIT

Type Letter	Conductor Size AWG,MCM	1/2	3/4	1	1-1/4	1-1/2	2	2-1/2	3	3-1/2	4	5	6
TW, T,	14	9	15	25	44	60	99	142					
RUH, RUW,	12	7	12	19	35	47	78	111	142				
XHHW	10	5	9	15	26	36	60	85	131	176			
(14 thru 8)	8	2	4	7	12	17	28	40	62	84	108		
RHW & RHH	14	6	10	16	29	40	65	93	143	192			
(no outer	12	4	8	13	24	32	53	76	117	157			
covering),	10	4	6	11	19	26	43	61	95	127	163		
THW	8	1	3	5	10	13	22	32	49	66	85	133	
TW,	6	1	2	4	7	10	16	23	36	48	62	97	141
T,	4	1	1	3	5	7	12	17	27	36	47	73	106
THW,	3	1	1	2	4	6	10	15	23	31	40	63	91
RUH	2	1	1	2	4	5	9	13	20	27	34	54	78
(6 thru 2)	1		1	1	3	4	6	9	14	19	25	39	57
RUW	0		1	1	2	3	5	8	12	16	21	33	49
(6 thru 2)	00		1	1	1	3	5	7	10	14	18	29	41
FEPB,	000		1	1	1	2	4	6	9	12	15	24	35
(6 thru 2)	0000			1	1	1	3	5	7	10	13	20	29
RHW & RHH	250			1	1	1	2	4	6	8	10	16	23
(no outer	300			1	1	1	2	3	5	7	9	14	20
covering)	350				1	1	1	3	4	6	8	12	18
	400				1	1	1	2	4	5	7	11	16
	500				1	1	1	1	3	4	6	9	14
	600					1	1	1	3	4	5	7	11
	700					1	1	1	2	3	4	7	10
	750					1	1	1	2	3	4	6	9
THWN,	14	13	24	39	69	94	154						
THHN,	12	10	18	29	51	70	114	164					
FEP,	10	6	11	18	32	44	73	104	160				
(14 thru 2),	8	3	5	9	16	22	36	51	79	106	136		
FEPB	6	1	4	6	11	15	26	37	57	76	96	154	
(14 thru 8),	4	1	2	4	7	9	16	22	35	47	60	94	137
PFA	3	1	1	3	6	8	13	19	29	39	51	80	116
(14 thru 4/0),	2	1	1	3	5	7	11	16	25	33	43	67	97
PFAH	1		1	1	3	5	8	12	18	25	32	50	72
(14 thru 4/0),	0		1	1	3	4	7	10	15	21	27	42	61
Z	00		1	1	2	3	6	8	13	17	22	35	51
(14 thru 4/0),	000			1	1	3	5	7	11	14	18	29	42
XHHW	0000			1	1	2	4	6	9	12	15	24	35
(4 thru 500MCM),	250				1	1	3	4	7	10	12	20	28
	300				1	1	3	4	6	8	11	17	24
	350				1	1	2	3	5	7	9	15	21
	400					1	1	3	5	6	8	13	19
	500					1	1	2	4	5	7	11	16
	600					1	1	1	3	4	5	9	13
	700						1	1	3	4	5	8	11
	750						1	1	2	3	4	7	11
XHHW	6	1	3	5	9	13	21	30	47	63	81	128	185
	600					1	1	1	3	4	5	9	13
	700					1	1	1	3	4	5	7	11
	750					1	1	1	2	3	4	7	10

EXPANSION CHARACTERISTICS OF PVC RIGID NONMETALLIC CONDUIT COEFFICIENT OF THERMAL EXPANSION = 3.38×10^5 in./in/°F

Temperature Change in Degrees F	Length Change in Inches per 100 ft. of PVC Conduit	Temperature Change in Degrees F	Length Change in Inches per 100 ft. of PVC Conduit
5	0.2	55	2.2
10	0.4	60	2.4
15	0.6	65	2.6
20	0.8	70	2.8
25	1.0	75	3.0
30	1.2	80	3.2
35	1.4	85	3.4
40	1.6	90	3.6
45	1.8	95	3.8
50	2.0	100	4.1
105	4.2	155	6.3
110	4.5	160	6.5
115	4.7	165	6.7
120	4.9	170	6.9
125	5.1	175	7.1
130	5.3	180	7.3
135	5.5	185	7.5
140	5.7	190	7.7
145	5.9	195	7.9
150	6.1	200	8.1

DESIGNED DIMENSIONS AND WEIGHTS OF RIGID STEEL CONDUIT

Nominal or Trade Size Of Conduit (Inches)	Inside Diameter (Inches)	Outside Diameter (Inches)	Wall Thickness (Inches)	Length Without Coupling Ft. & Ins.	Minimum Weight of Ten Unit Lengths with Couplings Attached (Pounds)
1/4	0.364	0.540	0.088	9–11-1/2	38.5
3/8	0.493	0.675	0.091	9–11-1/2	51.5
1/2	0.622	0.840	0.109	9–11-1/4	79.0
3/4	0.824	1.050	0.113	9–11-1/4	105.0
1	1.049	1.315	0.133	9–11	153.0
1-1/4	1.380	1.660	0.140	9–11	201.0
1-1/2	1.610	1.900	0.145	9–11	249.0
2	2.067	2.375	0.154	9–11	334.0
2-1/2	2.469	2.875	0.203	9–10-1/2	527.0
3	3.068	3.500	0.216	9–10-1/2	690.0
3-1/2	3.548	4.000	0.226	9–10-1/4	831.0
4	4.026	4.500	0.237	9–10-1/4	982.0
5	5.047	5.563	0.258	9-10	1344.0
6	6.065	6.625	0.280	9-10	1770.0

NOTE: The tolerances are:
Length: ± 1/4-inch (without coupling)
Outside Diameter: + 1/64-inch or –1/32-inch for the 1-1/2 inch and smaller sizes
± 1 percent for the 2-inch and larger sizes
Wall Thickness – 12-1/2 percent

DIMENSIONS OF THREADS FOR RIGID STEEL CONDUIT

Nominal or Trade Size of Conduit (Inches)	Threads per Inch	Pitch Diameter at End of Thread E_0 (Inches) Taper 3/4 Inch per Foot	Length of Thread (Inches)	
			Effective L_2	Over-All L_4
1/4	18	0.4774	0.40	0.59
3/8	18	0.6120	0.41	0.60
1/2	14	0.7584	0.53	0.78
3/4	14	0.9677	0.55	0.79
1	11-1/2	1.2136	0.68	0.98
1-1/4	11-1/2	1.5571	0.71	1.01
1-1/2	11-1/2	1.7961	0.72	1.03
2	11-1/2	2.2690	0.76	1.06
2-1/2	8	2.7195	1.14	1.57
3	8	3.3406	1.20	1.63
3-1/2	8	3.8375	1.25	1.68
4	8	4.3344	1.30	1.73
5	8	5.3907	1.41	1.84
6	8	6.4461	1.51	1.95

NOTE: The tolerances are:
Thread length (L_4, Column 5): ± 1 thread.
Pitch Diameter (Column 3): ± 1 turn is the maximum variation permitted from the gaging face of the working thread gages. This is equivalent to ± 1-1/2 turns from basic dimensions, since a variation of ± 1/2 turn from basic dimensions is permitted in working gages.

DESIGNED DIMENSIONS AND WEIGHTS OF COUPLINGS

Nominal or Trade Size of Conduit (Inches)	Outside Diameter (Inches)	Minimum Length (Inches)	Minimum Weight (Pounds)
1/4	0.719	1-3/16	0.055
3/8	0.875	1-3/16	0.075
1/2	1.010	1-9/16	0.115
3/4	1.250	1-5/8	0.170
1	1.525	2	0.300
1-1/4	1.869	2-1/16	0.370
1-1/2	2.155	2-1/16	0.515
2	2.650	2-1/8	0.671
2-1/2	3.250	3-1/8	1.675
3	3.870	3-1/4	2.085
3-1/2	4.500	3-3/8	3.400
4	4.875	3-1/2	2.839
5	6.000	3-3/4	4.462
6	7.200	4	7.282

NOTE: The tolerances are:
Outside Diameter: −1 percent for the 1-1/4-inch and larger sizes.
  −1/64-inch for sizes smaller than 1-1/4-inch.
No limit is placed on the plus tolerances given for this dimension.

DIMENSIONS OF 90-DEGREE ELBOWS AND WEIGHTS OF NIPPLES PER HUNDRED

Nominal or Trade Size of Conduit (Inches)	Elbows		Nipples	
	Minimum Radius to Center of Conduit (Inches)	Minimum Straight Length L₁ at Each End(Inches)	A	B
1/4	–	–	–	–
3/8	–	–	–	–
1/2	4	1-1/2	0.065	2
3/4	4-1/2	1-1/2	0.086	4
1	5-3/4	1-7/8	0.125	9
1-1/4	7-1/4	2	0.164	10
1-1/2	8-1/4	2	0.202	11
2	9-1/2	2	0.269	14
2-1/2	10-1/2	3	0.430	60
3	13	3-1/8	0.561	70
3-1/2	15	3-1/4	0.663	90
4	16	3-3/8	0.786	115
5	24	3-5/8	1.060	170
6	30	3-3/4	1.410	200

Each lot of 100 nipples shall weigh not less than the number of pounds determined by the formula:
$$W = 100\ LA - B$$
Where W = weight of 100 nipples in pounds   L = length of one nipple in inches
  A = weight of nipple per inch in pounds
  B = weight in pounds, lost in threading 100 nipples

DIMENSIONS AND WEIGHTS OF ELECTRICAL METALLIC TUBING

Nominal or Trade Size of Tubing (Inches)	Outside Diameter (Inches)	Minimum Wall Thickness (Inches)	Length (Feet)	Minimum Weight per 100 Feet (Pounds)
3/8	0.577	0.040	10	23
1/2	0.706	0.040	10	28.5
3/4	0.922	0.046	10	43.5
1	1.163	0.054	10	64
1-1/4	1.510	0.061	10	95
1-1/2	1.740	0.061	10	110
2	2.197	0.061	10	140

NOTE: The tolerances are:
Length: ± 1/4-inch
Outside Diameter: ± 0.005-inch
Wall Thickness: + 18 percent

DIMENSIONS OF 90-DEGREE ELBOWS

Nominal or Trade Size of Tubing (Inches)	Minimum Radius to Center of Tubing (Inches)	Minimum Straight Length Ls at Each End (Inches)
1/2	4	1-1/2
3/4	4-1/2	1-1/2
1	5-3/4	1-7/8
1-1/4	7-1/4	2
1-1/2	8-1/4	2
2	9-1/2	2

APPROXIMATE SPACING OF CONDUIT BUSHINGS, CHASE NIPPLES AND LOCK NUTS

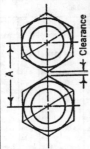

"D" in Simplest Practical Dimension	
Conduit	"D"
1/2	1-1/8
3/4	1-3/8
1	1-5/8
1-1/4	2
2	2-7/8
2-1/2	3-1/2
3	4-5/16
3-1/2	4-7/8

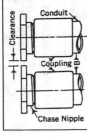

Size of Conduit		Clearance = 1/8"								
		1/2"	3/4'	1"	1-1/4"	1-1/2"	2"	2-1/2"	3"	3-1/2"
1/2"	A	1-1/4	1-3/8	1-1/2	1-11/16	1-13/16	2-1/8	2-7/16	2-3/4	3-1/8
	B	.41	.43	.42	.43	.44	.52	.58	.58	.70
3/4"	A	1-3/8	1-1/2	1-5/8	1-13/16	1-15/16	2-1/4	2-9/16	3	3-1/4
	B	.43	.45	.44	.46	.46	.54	.60	60	.72
1"	A	1-1/2	1-5/8	1-3/4	1-15/16	2-1/16	2-3/8	2-11/16	3	3-3/8
	B	.43	.45	.44	.46	.46	.54	.60	.60	.73
1-1/4"	A	1-11/16	1-13/16	1-15/16	2-1/8	2-1/4	2-9/16	2-7/8	3-1/4	3-9/16
	B	.43	.46	.46	.46	.47	.55	.62	.67	.74
1-1/2"	A	1-13/16	1-15/16	2-1/16	2-1/4	2-3/8	2-11/16	3	3-3/8	3-11/16
	B	.44	.46	.46.	.47	.47	.56	.62	.67	.74
2"	A	2-1/8	2-1/4	2-3/8	2-9/16	2-11/16	3	3-5/16	3-5/8	4
	B	.52	.54	.53	.55	.56	.63	.63	.69	.82
2-1/2"	A	2-7/16	2-9/16	2-11/16	2-7/8	3	3-5/16	3-5/8	3-15/16	4-5/16
	B	.58	.60	.60	.61	.62	.69	.76	.76	.89
3"	A	2-3/4	3	3	3-1/4	3-3/8	3-5/8	3-15/16	4-7/16	4-3/4
	B	.58	.60	.60	.67	.67	.69	.74	.94	1.00
3-1/2"	A	3-1/8	3-1/4	3-3/8	3-9/16	3-11/16	4	4-5/16	4-3/4	5
	B	.70	.72	.73	.74	.74	.82	.89	1.00	1.00

Size of Conduit		Clearance = 1/4"								
		1/2"	3/4'	1"	1-1/4"	1-1/2"	2"	2-1/2"	3"	3-1/2"
1/2"	A	1-3/8	1-1/2	1-5/8	1-13/16	1-15/16	2-1/4	2-9/16	2-7/8	3-1/4
	B	.53	.55	.54	.55	.56	.64	.70	.70	.83
3/4"	A	1-1/2	1-5/8	1-3/4	1-15/16	2-1/16	2-3/8	2-11/16	3	3-3/8
	B	.55	.57	.56	.58	.58	.66	.72	.72	.86
1"	A	1-5/8	1-3/4	1-7/8	2-1/16	2-3/16	2-1/2	2-13/16	3-1/8	3-1/2
	B	.55	.57	.56	.58	.58	.66	.72	.72	.85
1-1/4"	A	1-13/16	1-15/16	2-1/16	2-1/4	2-3/8	2-11/16	3	3-3/8	3-11/16
	B	.55	.58	.58	.58	.59	.67	.73	.79	.86
1-1/2"	A	1-15/16	2-1/16	2-3/16	2-3/8	2-1/2	2-13/16	3-1/8	3-1/2	3-13/16
	B	.56	.58	.58	.59	.59	.68	.74	.79	.87
2"	A	2-1/4	2-3/8	2-1/2	2-11/16	2-13/16	3-1/8	3-7/16	3-3/4	4-1/8
	B	.64	.66	.65	.67	.68	.75	.81	.81	.94
2-1/2"	A	2-9/16	2-11/16	2-13/16	3	3-1/8	3-7/16	3-3/4	4-1/16	4-7/16
	B	.70	.72	.72	.73	.74	.81	.87	.88	1.00
3"	A	2-7/8	3	3-1/8	3-3/8	3-1/2	3-3/4	4-1/16	4-9/16	4-7/8
	B	.70	.72	.72	.79	.79	.81	.88	1.06	1.12
3-1/2"	A	3-1/4	3-3/8	3-1/2	3-11/16	3-13/16	4-1/8	4-7/16	4-7/8	5-1/8
	B	.83	.86	.85	.86	.87	.94	1.00	1.12	1.12

APPROXIMATE SPACING OF CONDUIT BUSHINGS, CHASE NIPPLES AND LOCK NUTS

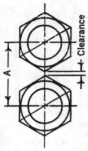

Size of Conduit		Clearance = 3/8"								
		1/2"	3/4"	1"	1-1/4"	1-1/2"	2"	2-1/2"	3"	3-1/2"
1/2"	A	1-1/2	1-5/8	1-3/4	1-15/16	2-1/16	2-3/8	2-11/16	3	3-3/8
	B	.66	.68	.67	.68	.69	.77	.83	.83	.96
3/4"	A	1-5/8	1-3/4	1-7/8	2-1/16	2-3/16	2-1/2	2-13/16	3-1/8	3-1/2
	B	.68	.70	.69	.71	.71	.79	.85	.85	.98
1"	A	1-3/4	1-7/8	2	2-3/16	2-5/16	2-5/8	2-15/16	3-1/4	3-5/8
	B	.68	.69	.69	.71	.71	.78	.85	.85	.98
1-1/4"	A	1-15/16	2-1/16	2-3/16	2-3/8	2-1/2	2-13/16	3-1/8	3-1/2	3-13/16
	B	.68	.71	.71	.71	.72	.80	.87	.92	.99
1-1/2"	A	2-1/16	2-3/16	2-5/16	2-1/2	2-5/8	2-15/16	3-1/4	3-5/8	3-15/16
	B	.69	.71	.71	.72	.72	.81	.87	.92	.99
2"	A	2-3/8	2-1/2	2-5/8	2-13/16	2-15/16	3-1/4	3-7/16	3-7/8	4-1/4
	B	.77	.79	.78	.80	.81	.88	.88	.94	1.07
2-1/2"	A	2-11/16	2-13/16	2-15/16	3-1/8	3-1/4	3-9/16	3-7/8	4-3/16	4-7/16
	B	.83	.85	.85	.86	.87	.94	1.01	1.01	1.13
3"	A	3	3-1/8	3-1/4	3-1/2	3-5/8	3-7/8	4-3/16	4-11/16	5
	B	.83	.85	.85	.92	.92	.94	.99	.99	1.25
3-1/2"	A	3-3/8	3-1/2	3-5/8	3-13/16	3-15/16	4-1/4	4-9/16	5	5-1/4
	B	.96	.98	.98	.99	.99	1.07	1.13	1.25	1.25

"D" in Simplest Practical Dimension

Conduit	"D"
1/2	1-1/8
3/4	1-3/8
1	1-5/8
1-1/4	2
1-1/2	2-1/4
2	2-7/8
2-1/2	3-1/2
3	4-5/16
3-1/2	4-7/8

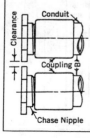

Size of Conduit		Clearance = 1/2"								
		1/2"	3/4"	1"	1-1/4"	1-1/2"	2"	2-1/2"	3"	3-1/2"
1/2"	A	1-5/8	1-3/4	1-7/8	2-1/16	2-3/16	2-1/2	2-13/16	3-1/8	3-1/2
	B	.78	.80	.79	.80	.81	.89	.95	.95	1.08
3/4"	A	1-3/4	1-7/8	2	2-3/16	2-5/16	2-5/8	2-15/16	3-1/4	3-5/8
	B	.80	.82	.81	.83	.83	.91	.97	.97	1.10
1"	A	1-7/8	2	2-1/8	2-5/16	2-7/16	2-3/4	3-1/16	3-3/8	3-3/4
	B	.80	.81	.81	.83	.83	.90	97	.97	1.10
1-1/4"	A	2-1/16	2-3/16	2-5/16	2-1/2	2-5/8	2-15/16	3-1/4	3-5/8	3-15/16
	B	.80	.83	.83	.83	.84	.92	.99	1.04	1.11
1-1/2"	A	2-3/16	2-5/16	2-7/16	2-5/8	2-3/4	3-1/16	3-3/8	3-3/4	4-1/16
	B	.81	.83	.83	.84	.84	.93	.98	1.04	1.11
2"	A	2-1/2	2-5/8	2-3/4	2-15/16	3-1/16	3-3/8	3-11/16	4	4-3/8
	B	.89	.91	.90	.92	.93	1.00	1.00	1.06	1.20
2-1/2"	A	2-13/16	2-15/16	3-1/16	3-1/4	3-3/8	3-11/16	4	4-5/16	4-11/16
	B	.95	.97	.97	.98	.99	1.06	1.13	1.13	1.26
3"	A	3-1/8	3-1/4	3-3/8	3-5/8	3-3/4	4	4-5/16	4-13/16	5-1/8
	B	.95	.97	.97	1.04	1.04	1.06	1.14	1.14	1.31
3-1/2"	A	3-1/2	3-5/8	3-3/4	3-15/16	4-1/16	4-1/8	4-11/16	5-1/16	5-3/8
	B	1.01	1.04	1.03	1.04	1.11	1.13	1.31	1.31	1.37

APPROXIMATE SIZES OF CONDUITS, COUPLINGS, CHASE NIPPLES AND BUSHINGS

Coupling

Conduit

Chase Nipples

Conduit Bushings

Size of Conduit	A	B	C	D	E	F	G	H	I	J
1/2"	14.0	.82	.85	.62	1.00	1.15	.62	.12	.50	.62
3/4"	14.0	1.02	1.12	.82	1.25	1.44	.81	.19	.62	.82
1"	11.5	1.28	1.67	1.04	1.37	1.59	.94	.25	.69	1.04
1-1/4"	11.5	1.63	2.24	1.38	1.75	2.02	1.06	.25	.81	1.38
1-1/2"	11.5	1.87	2.68	1.61	2.00	2.31	1.12	.31	.81	1.61
2"	11.5	2.34	3.61	2.06	2.50	2.89	1.31	.31	1.00	2.06
2-1/2"	8.0	2.82	5.74	2.46	3.00	3.46	1.44	.37	1.06	2.46
3"	8.0	3.44	7.54	3.06	3.75	4.33	1.50	.37	1.12	3.06
3-1/2"	8.0	3.94	9.00	3.54	4.25	4.91	1.62	.44	1.19	3.54

Size of Conduit	K	L	M	N	O	P	Q	R	S	T
1/2"	.84	.62	1.00	.94	.37	.12	.19	.06	1.37	1.12
3/4"	1.05	.75	1.25	1.12	.44	.12	.25	.06	1.56	1.31
1"	1.31	1.00	1.50	1.37	.50	.16	.25	.09	1.75	1.62
1-1/4"	1.66	1.25	1.81	1.75	.56	.19	.28	.09	2.12	2.00
1-1/2"	1.90	1.50	2.12	2.00	.56	.19	.28	.09	2.50	2.25
2"	2.37	1.94	2.56	2.37	.62	.19	.31	.12	2.62	2.75
2-1/2"	2.87	2.37	3.06	2.87	.75	.25	.37	.12	2.87	3.31
3"	3.50	2.87	3.75	3.50	.81	.25	.37	.19	3.06	3.93
3-1/2"	4.00	3.25	4.25	4.00	1.00	.37	.44	.19	3.62	4.43

3-25

COMPARATIVE WEIGHTS OF COPPER AND ALUMINUM CONDUCTORS / Lbs. per 1,000 Ft.

Size AWG or MCM	Bare - Solid			Bare - Stranded		
	Cu.	Al.	Diff.	Cu.	Al.	Diff.
18	4.92	1.49	3.43	5.02	1.53	3.49
16	7.82	2.38	5.44	7.97	2.43	5.54
14	12.43	3.78	8.65	12.68	3.86	8.82
12	19.77	6.01	13.76	20.16	6.13	14.03
10	31.43	9.55	21.88	32.06	9.75	22.31
8	50.0	15.2	34.8	51.0	15.5	35.5
6	79.5	24.2	55.3	81.0	24.6	56.4
4	126.4	38.4	88.0	128.9	39.2	89.7
2	200.9	60.8	140.1	204.9	62.3	142.6
1	253.3	77.0	176.3	258.4	78.6	179.8
1/0	319.5	97.2	222.3	325.8	99.1	226.7
2/0	402.8	122.6	280.2	410.9	124.9	286.0
3/0	507.9	154.6	353.3	518.1	157.5	360.6
4/0	640.5	194.9	445.6	653.3	198.6	454.7
250				771.9	234.7	537.2
300				926.3	281.6	644.7
350				1081.	328.6	752.
400				1235.	375.5	859.
450				1389.	422.4	967.
500				1544.	469.4	1075.
600				1853.	563.	1290.
700				2161.	657.	1504.
750				2316.	704.	1612.
800				2470.	751.	1719.
900				2779.	845.	1934.
1000				3088.	939.	2149.
1250				3859.	1173.	2686.
1500				4631.	1410.	3221.
1750				5403.	1645.	3758.
2000				6175.	1880.	4295.

VOLTAGE DROP TABLE

Conductor Size	DC	Volts Drop Per 1000 Ampere-Feet						
		AC System						
		LOAD POWER FACTOR in Percent						
		100	95	90	85	80	75	70

For DC circuit or single phase, 60 cycle, 2 wire system or 3 wire system with balanced load. Copper conductors, 70°C copper temperature. 600 V class single-conductor cables in steel conduit.

Conductor Size	DC	100	95	90	85	80	75	70
14	6.29	6.29	6.06	5.78	5.54	5.26	4.97	4.74
12	3.93	3.93	3.81	3.64	3.46	3.29	3.13	2.95
10	2.48	2.48	2.44	2.31	2.19	2.08	1.96	1.85
8	1.56	1.56	1.51	1.47	1.41	1.34	1.27	1.20
6	0.999	1.011	0.987	0.953	0.918	0.872	0.826	0.774
4	0.631	0.635	0.641	0.624	0.600	0.578	0.554	0.528
2	0.396	0.404	0.418	0.413	0.400	0.386	0.372	0.358
1	0.314	0.323	0.356	0.337	0.330	0.322	0.311	0.300
1/0	0.249	0.263	0.280	0.282	0.277	0.269	0.263	0.255
2/0	0.198	0.214	0.233	0.236	0.233	0.230	0.226	0.222
3/0	0.157	0.173	0.196	0.206	0.204	0.200	0.194	0.188
4/0	0.123	0.141	0.163	0.170	0.171	0.170	0.169	0.166
250 MCM	0.1041	0.121	0.146	0.152	0.155	0.155	0.155	0.154
300 MCM	0.0870	0.1040	0.128	0.135	0.139	0.140	0.141	0.141
350 MCM	0.0746	0.0912	0.117	0.125	0.128	0.131	0.131	0.131
400 MCM	0.0652	0.0855	0.1086	0.117	0.120	0.122	0.124	0.125
500 MCM	0.0528	0.0733	0.0959	0.1040	0.1086	0.111	0.113	0.114
750 MCM	0.0347	0.0589	0.0808	0.0884	0.0940	0.0976	0.0999	0.1020

For 3-phase, 60 cycle, 3-wire or 4-wire balanced system. Copper conductors, 70°C copper temperature, 600 V. class single-conductor cables in steel conduit.

Conductor Size	DC	100	95	90	85	80	75	70
14		5.45	5.25	5.00	4.80	4.55	4.30	4.10
12		3.40	3.30	3.15	3.00	2.85	2.70	2.55
10		2.15	2.10	2.00	1.90	1.80	1.70	1.60
8		1.35	1.31	1.27	1.22	1.16	1.10	1.04
6		0.875	0.855	0.825	0.795	0.755	0.715	0.670
4		0.550	0.555	0.540	0.520	0.500	0.480	0.457
2		0.350	0.362	0.358	0.346	0.334	0.322	0.310
1		0.280	0.308	0.292	0.286	0.279	0.269	0.260
1/0		0.228	0.242	0.244	0.240	0.233	0.228	0.221
2/0		0.185	0.202	0.204	0.202	0.199	0.196	0.192
3/0		0.150	0.170	0.178	0.177	0.173	0.168	0.163
4/0		0.122	0.141	0.147	0.148	0.147	0.146	0.144
250 MCM		0.105	0.126	0.132	0.134	0.134	0.134	0.133
300 MCM		0.0900	0.111	0.117	0.120	0.121	0.122	0.122
350 MCM		0.0790	0.101	0.108	0.111	0.113	0.114	0.114
400 MCM		0.0740	0.0940	0.101	0.104	0.106	0.107	0.108
500 MCM		0.0635	0.0630	0.0900	0.0940	0.0964	0.0974	0.0988
750 MCM		0.0510	0.0700	0.0765	0.0814	0.0845	0.0865	0.0883

NOTE: Length to be used is the distance from point of supply to load, not amount of wire in circuit.

WIRE DATA
STANDARD STRANDED CONDUCTORS

Size C.M.	Number of Wires in the Strand								
	7	19	37	7x7–49	61	91	127	169	217
	Diameter in Inches of Each Wire in the Strand								
2000000	.5345	.3243	.2325	.202	.181	.1482	.1255	.1088	.096
1750000	.5000	.3034	.2175	.189	.1694	.1386	.1157	.1003	.0898
1500000	.4629	.2810	.2013	.175	.1568	.1284	.1087	.0942	.0831
1250000	.4226	.2565	.1838	.1507	.1431	.1172	.0992	.086	.0759
1000000	.378	.2294	.1644	.1429	.1285	.1048	.0887	.0769	.0678
950000	.3684	.2236	.1602	.1392	.1248	.1021	.0864	.075	.0662
900000	.3586	.2176	.1559	.1355	.1215	.0994	.0841	.073	.0644
850000	.3484	.2115	.1516	.1317	.1181	.0966	.0818	.0709	.0626
800000	.338	.205	.147	.1278	.1145	.0937	.0793	.0687	.0607
750000	.3273	.1986	.1424	.1237	.1109	.0908	.0768	.0666	.0588
700000	.3163	.1919	.1375	.1195	.1071	.0883	.0742	.0644	.0568
650000	.3047	.1850	.1325	.1152	.1032	.0845	.0716	.0620	.0547
600000	.2928	.1778	.1273	.1107	.0992	.0812	.0687	.0596	.0526
550000	.2803	.1701	.1219	.106	.0950	.0777	.0658	.0570	.0503
500000	.2672	.1622	.1162	.101	.0905	.0741	.0627	.0544	.048
450000	.2535	.1539	.1103	.0958	.0859	.0703	.0595	.0516	.0455
400000	.2391	.1451	.1040	.0904	.081	.0663	.0561	.0487	.0429
350000	.2236	.1357	.0973	.0845	.0757	.0620	.0526	.0455	.0401
300000	.207	.1257	.0901	.0783	.0701	.0573	.0486	.0421	.0372
250000	.189	.1147	.0822	.0714	.064	.0524	.0444	.0384	.0340

Size B & S									
0000	.1736	.1055	.0756	.0657	.0589	.0482	.0408	–	–
000	.1548	.0940	.0673	.0586	.0525	.0429	.0363	–	–
00	.1378	.0836	.0599	.0521	.0467	.0382	.0323	–	–
0	.1228	.0746	.0534	.0464	.0416	.0340	.0288	–	–
1	.1093	.0663	.0475	.0413	.0370	.0303	.0252	–	–
2	.0973	.0592	.0423	.0369	.0329	.0269	.0228	–	–
3	.0867	.0526	.0377	.0327	.0294	.0240	.0203	–	–
4	.0772	.0468	.0335	.0291	.0261	.0214	.0179	–	–
5	.0687	.0417	.0299	.026	.0233	.0190	.0161	–	–
6	.0612	.0372	.0266	.0231	.0207	.0169	.0143	–	–
7	.0545	.0331	.0237	–	.0184	.0151	.0128	–	–
8	.0485	.0293	.0211	.0184	.0164	.0135	.0114	–	–
9	.0435	.0262	.0187	–	.0146	.0120	.0101	–	–
10	.0385	.0223	.0168	–	.0129	.0106	.0090	–	–

VOLTAGE DROP AMPERE-FEET
Copper Conductors, 70°C Copper Temp.,
600 V. Class Single Conductor Cables in Steel Conduit

Maximum circuit ampere-feet without exceeding specified percentage voltage drop, various circuit voltages and power factors.

Conductor Size AWG or MCM	DC Circuits 1% Drop on 120 V	60 Cycle AC Circuits				
		1% Drop, 1.00 P.F.		3% Drop, 0.85 P.F.		
		120 V 1-Phase	208 V 3-Phase	115 V 1-Phase	208 V 3-Phase	220 V 3-Phase
14	191	191	382	623	1,300	1.380
12	305	305	612	998	2,080	2,200
10	484	484	968	1,580	3,280	3,470
8	770	770	1,540	2,450	5,110	5,410
6	1,200	1,190	2,380	3,800	7,850	8,310
4	1,900	1,890	3,780	5,750	12,000	12,700
2	3,030	2,970	5,950	8,620	18,000	19,100
1	3,820	3,710	7,430	10,400	21,800	23,100
1/0	4,820	4,560	9,120	12,500	26,000	27,500
2/0	6,060	5,610	11,200	14,800	30,800	32,700
3/0	7,650	6,940	13,900	16,900	35,200	37,100
4/0	9,760	8,520	17,100	20,200	42,100	44,600
250	11,500	9,930	19,800	22,300	46,500	49,300
300	13,800	11,500	23,100	24,800	52,000	55,000
350	16,100	13,200	26,300	27,000	56,200	59,500
400	18,400	14,000	28,100	28,700	60,000	63,500
500	22,700	16,400	32,800	31,800	66,400	70,300
750	34,600	20,400	40,800	36,700	76,600	81,200

NOTE: Length to be used is the distance from point of supply to load, not amount of wire in circuit.

MELTING POINT AND RELATIVE CONDUCTIVITY OF DIFFERENT METALS AND ALLOYS

Metals	Relative Conductivity	Melting Point °F
Pure silver	106.0	1760
Pure copper	100.0	1980
Refined and crystalized copper	99.9	—
Telegraphic silicious bronze	98.0	—
Alloy of copper and silver (50%)	86.65	—
Silicide of copper, 4% Si	75.0	—
Pure gold	71.3	1950
Pure aluminum	64.5	1220
Silicide of copper, 12% Si	54.7	—
Tin with 12% of sodium	46.9	—
Telephonic silicious bronze	35.0	—
Copper with 10% of lead	30.0	—
Pure zinc	29.2	790
Telephonic phosphor-bronze	29.0	—
Silicious brass, 25% zinc	26.4	—
Brass with 35% zinc	21.59	—
Phosphor-tin	17.7	—
Alloy of gold and silver (50%)	16.12	—
Swedish iron	16.4	2800
Pure platinum	16.3	3230
Pure banca tin	15.5	442
Antimonial copper	12.7	—
Aluminum bronze (10%)	12.6	—
Siemens Steel	12.0	—
Copper with 10% of nickel	10.6	—
Cadmium amalgam (15%)	10.2	—
Dronier mercurial bronze	10.14	—
Arsenical copper (10%)	9.1	—
Bronze with 20% tin	8.4	—
Pure lead	7.8	620
Phosphor-bronze, 10% tin	6.5	—
Pure nickel	5.8	2640
Phosphor-copper, 9% phos.	4.9	—
Antimony	4.4	1167

COPPER BUS-BAR DATA
Sizes, Weights and Resistances

Thickness inch	Width inch	Wts. per Ft. at .3213 Lbs. per Cubic in.	Area in Square In.	Ohms per Ft. at 8.341 per Sq. Mil. Ft.	Capacity in Ampere
1/16	1/2	.1205	.0313	.00026691	30
1/16	3/4	.1807	.0469	.00017790	50
1/16	1	.2410	.0625	.00013344	60
1/16	1-1/2	.3615	.0938	.00008897	90
1/8	1/2	.2410	.0625	.00013344	75
1/8	3/4	.3615	.0938	.00008897	90
1/8	1	.4820	.125	.00006672	125
1/8	1-1/2	.7230	.1875	.00004448	200
1/8	2	.9640	.25	.00003336	250
1/4	3/4	.7230	.1875	.00004448	185
1/4	1	.9640	.25	.00003336	250
1/4	1-1/4	1.205	.3125	.00002669	315
1/4	1-1/2	1.446	.375	.00002224	375
1/4	1-3/4	1.687	.4375	.00001906	435
1/4	2	1.928	.5	.00001668	500
1/4	2-1/4	2.169	.5625	.00001482	565
1/4	2-1/2	2.410	.625	.00001334	630
1/2	3/4	1.446	.375	.00002224	370
1/2	1	1.928	.500	.00001668	500
1/2	1-1/4	2.410	.625	.00001334	625
1/2	1-1/2	2.892	.750	.00001112	750
1/2	1-3/4	3.374	.875	.00000953	875
1/2	2	3.856	1.	.00000834	1000
1/2	2-1/4	4.338	1.125	.00000741	1185
1/2	2-1/2	4.820	1.25	.00000667	1250
1/2	2-3/4	5.304	1.375	.00000606	1375
1/2	3	5.784	1.500	.00000556	1500
1/2	3-1/4	6.266	1.625	.00000513	1625
1/2	3-1/2	6.748	1.750	.00000475	1750
1/2	3-3/4	7.23	1.875	.00000444	1875
1/2	4	7.712	2.000	.00000417	2000
3/4	1	2.892	.750	.00001112	750
3/4	1-1/2	4.338	1.125	.00000741	1125
3/4	2	5.784	1.500	.00000556	1500
3/4	2-1/2	7.23	1.875	.00000444	1875
3/4	3	8.676	2.250	.00000370	2250
3/4	3-1/2	10.122	2.625	.00000317	2650
3/4	4	11.568	3.000	.00000278	3000

Carrying capacity is figured at 1,000 amperes per square inch.

WIRE AND SHEET METAL GAUGES
In Approximate Decimals of an Inch

Gauge Numbers	United States	American or Brown & Sharpe	Washburn & Moen. Am. Steel & Wire Co. Roebling	Trenton Iron Co.	Birmingham or Stubs' Iron Wire	Stubs' Steel Wire	British Imperial	Gauge Numbers
7 - 0	.500	———	.4900	———			.500	7 - 0
6 - 0	.469	———	.4615				.464	6 - 0
5 - 0	.438	———	.4305	.450			.432	5 - 0
4 - 0	.406	.460	.3938	.400	.454		.400	4 - 0
000	.375	.410	.3625	.360	.425		.372	000
00	.344	.365	.3310	.330	.380		.348	00
0	.313	.325	.3065	.305	.340		.324	0
1	.281	.289	.2830	.285	.300	.227	.300	1
2	.266	.258	.2625	.265	.284	.219	.276	2
3	.250	.229	.2437	.245	.259	.212	.252	3
4	.234	.204	.2253	.225	.238	.207	.232	4
5	.219	.182	.2070	.205	.220	.204	.212	5
6	.203	.162	.1920	.190	.203	.201	.192	6
7	.188	.144	.1770	.175	.180	.199	.176	7
8	.172	.128	.1620	.160	.165	.197	.160	8
9	.156	.114	.1483	.145	.148	.194	.144	9
10	.141	.102	.1350	.130	.134	.191	.128	10
11	.125	.0907	.1205	.1175	.120	.188	.116	11
12	.109	.0808	.1055	.105	.109	.185	.104	12
13	.0938	.0720	.0915	.0925	.095	.182	.092	13
14	.0781	.0641	.0800	.0806	.083	.180	.080	14
15	.0703	.0571	.0720	.070	.072	.178	.072	15
16	.0625	.0508	.0625	.061	.065	.175	.064	16
17	.0563	.0453	.0540	.0525	.058	.172	.056	17
18	.0500	.0403	.0475	.045	.049	.168	.048	18
19	.0438	.0359	.0410	.040	.042	.164	.040	19
20	.0375	.0320	.0348	.035	.035	.161	.036	20
21	.0344	.0285	.0318	.031	.032	.157	.032	21
22	.0313	.0253	.0286	.028	.028	.155	.028	22
23	.0281	.0226	.0258	.025	.025	.153	.024	23
24	.0250	.0201	.0230	.0225	.022	.151	.022	24
25	.0219	.0179	.0204	.020	.020	.148	.020	25
26	.0188	.0159	.0181	.018	.018	.146	.018	26
27	.0172	.0142	.0173	.017	.016	.143	.0164	27
28	.0156	.0126	.0162	.016	.014	.139	.0149	28
29	.0141	.0113	.0150	.015	.013	.134	.0136	29
30	.0125	.0100	.0140	.014	.012	.127	.0124	30
31	.0109	.0089	.0132	.013	.010	.120	.0116	31
32	.0102	.0080	.0128	.012	.009	.115	.0108	32
33	.0094	.0071	.0118	.011	.008	.112	.010	33
34	.0086	.0063	.0104	.010	.007	.110	.0092	34
35	.0078	.0056	.0095	.0095	.005	.108	.0084	35
36	.0070	.0050	.0090	.009	.004	.106	.00076	36
37	.0066	.0045	.0085	.0085		.103	.0068	37
38	.0063	.0040	.0080	.008	———	.101	.006	38
39	.0059	.0035	.0075	.0075	———	.099	.0052	39
40	.0055	.0031	.0070	.007	———	.097	.0048	40

PLASTIC PIPE

Although there are many plasitc pipe types listed below, PVC and ABS are by far the most common types. It is imperative that the correct primers and solvents be used on each type of pipe or the joints will not seal properly and the overall strength will be weakened.

Types of Plastic Pipe

Type	Characteristics
PVC	Polyvinyl Chloride, Type 1, Grade 1. This pipe is strong, rigid and resistant to a variety of acids and bases. Some solvents and chlorinated hydrocarbons may damage the pipe. PVC is very common, easy to work with and readily available at most hardware stores. Maximum useable temperatures is 140°F (60°C) and pressure ratings start at a minimum of 125 to 200 psi (check for specific ratings on the pipe or ask the seller.) PVC can be used with water, gas, and drainage systems but NOT with hot water systems.
ABS	Acrylonitile Butadiene Styrene, Type 1. This pipe is strong, rigid and resistant to a variety of acids and bases. Some solvents and chlorinated hydrocarbons may damage the pipe. ABS is very common, easy to work with and readily available at most hardware stores. Maximum useable temperature is 160°F (71°C) at low pressures it is most common as a DWV pipe.
CPVC	Chlorinated polyvinyl chloride. Similar to PVC but designed specifically for piping water at up to 180°F (82°C) (can actually withstand 200°F for a limited time). Pressure rating is 100 psi.
PE	Polyethylene. A flexible pipe for pressurized water systems such as sprinklers. Not for hot water.
PB	Polybutylene. A flexible pipe for pressurized water both hot and cold. ONLY compression and banded type joints can be used.
Polypropylene	**Low pressure**, lightweight material that is good up to 180°F (82°C). Highly resistant to acids, bases, and many solvents. Good for laboratory plumbing.
PVDF	Polyvinylidene fluoride. Strong, very tough, and resistant to abrasion, acids, bases, solvents, and much more. Good to 280°F (138°C). Good in lab.
FRP Epoxy	A thermosetting plastic over fiberglass. Very high strength and excellent chemical resistance. Good to 220°F (105°C). Excellent for labs.

PLASTIC PIPE

Nominal Size Inches	Actual OD Inches	PVC Sched. 40 Wall Th. Inch	PVC Sched. 40 Weight Lbs/Foot	PVC Sched. 80 Wall Th. Inch	PVC Sched. 80 Weight Lbs/Foot
1/4	0.540			0.119	0.10
1/2	0.840	0.109	0.16	0.147	0.21
3/4	1.050	0.113	0.22	0.154	0.28
1	1.315	0.133	0.32	0.179	0.40
1-1/4	1.660	0.140	0.43	0.191	0.57
1-1/2	1.900	0.145	0.52	0.200	0.69
2	2.375	0.154	0.70	0.218	0.95
2-1/2	2.875	0.203	1.10	0.276	1.45
3	3.500	0.216	1.44	0.300	1.94
4	4.500	0.237	2.05	0.337	2.83
6	6.625	0.280	3.61	0.432	5.41
8	8.625	0.322	5.45	0.500	8.22
10	10.750	0.365	7.91	0.593	12.28
12	12.750	0.406	10.35	0.687	17.10

Nominal Size Inches	Actual OD Inches	CPVC Sched. 40 Wall Th. Inch	CPVC Sched. 40 Weight Lbs/Foot	CPVC Sched. 80 Wall Th. Inch	CPVC Sched. 80 Weight Lbs/Foot
1/4	0.540			0.119	0.12
1/2	0.840	0.109	0.19	0.147	0.24
3/4	1.050	0.113	0.25	0.154	0.33
1	1.315	0.133	0.38	0.179	0.49
1-1/4	1.660	0.140	0.51	0.191	0.67
1-1/2	1.900	0.145	0.61	0.200	0.81
2	2.375	0.154	0.82	0.218	1.09
2-1/2	2.875	0.203	1.29	0.276	1.65
3	3.500	0.216	1.69	0.300	2.21
4	4.500	0.237	2.33	0.337	3.23
6	6.625	0.280	4.10	0.432	6.17
8	8.625			0.500	9.06

Nominal Size Inches	Actual OD Inches	PVDF Sched. 80 Wall Th. Inch	PVDF Sched. 80 Weight Lbs/Foot	Polypropylene 80 Wall Th. Inch	Polypropylene 80 Weight Lbs/Foot
1/2	0.840	0.147	0.24	0.147	0.14
3/4	1.050	0.154	0.33	0.154	0.19
1	1.315	0.179	0.49	0.179	0.27
1-1/4	1.660	0.191	——	0.191	0.38
1-1/2	1.900	0.200	0.81	0.200	0.45
2	2.375	0.218	1.13	0.218	0.62

Pipe Schedule Number = 1000 X $\dfrac{\text{psi internal pressure}}{\text{psi allowable fiber stress}}$

CHAPTER 4
COMMUNICATIONS

COMMON TELEPHONE CONNECTIONS

The most common and simplest type of communication installation is the single line telephone. The typical telephone cable (sometimes called quad cable) contains four wires, colored green, red, black, and yellow. A one line telephone requires only two wires to operate. In almost all circumstances, green and red are the two conductors used. In a common four-wire modular connector, the green and red conductors are found in the inside positions, with the black and yellow wires in the outer positions.

As long as the two center conductors of the jack (again, always green and red) are connected to live phone lines, the telephone should operate.

Two-line phones generally use the same four wire cables and jacks. In this case, however, the inside two wires (green and red) carry line 1, and the outside two wires (black and yellow) carry line 2.

COLOR-CODING OF CABLES

The color coding of twisted-pair cable uses a color pattern that identifies not only what conductors make up a pair but also what pair in the sequence it is, relative to other pairs within a multipair sheath. This is also used to determine which conductor in a pair is the *tip* conductor and which is the *ring* conductor. (The tip conductor is the positive conductor, and the ring conductor is the negative conductor.)

The banding scheme uses two opposing colors to represent a single pair. One color is considered the primary while the other color is considered the secondary. For example, given the primary color of white and the secondary color of blue, a single twisted-pair would consist of one cable that is white with blue bands on it. The five primary colors are white, red, black, yellow, and violet.

In multi-pair cables the primary color is responsible for an entire group of pairs (five pairs total). For example, the first five pairs all have the primary color of white. Each of the secondary colors, blue, orange, green, brown, and slate, are paired in a banded fashion with white. This continues through the entire primary color scheme for all four primary colors (comprising 25 individual pairs). In larger cables (50 pairs and up), each 25-pair group is wrapped in a pair of ribbons, again representing the groups of primary colors matched with their respective secondary colors. These color coded band markings help cable technicians to quickly identify and properly terminate cable pairs.

EIA COLOR CODE

You should note that the new EIA color code calls for the following color coding:

Pair 1	–	White/Blue (white with blue stripe) and Blue
Pair 2	–	White/Orange and Orange
Pair 3	–	White/Green and Green
Pair 4	–	White/Brown and Brown

TWISTED-PAIR PLUGS AND JACKS

One of the more important factors regarding twisted-pair implementations is the cable jack or cross-connect block. These items are vital since without the proper interface, any twisted-pair cable would be relatively useless. In the twisted-pair arena, there are three major types of twisted-pair jacks:

RJ-type connectors (phone plugs)

Pin-connector

Genderless connectors (IBM sexless data connectors)

The RJ-type (registered jack) name generally refers to the standard format used for most telephone jacks. The term pin-connector refers to twisted pair connectors, such as the RS-232 connector, which provide connection through male and female pin receptacles. Genderless connectors are connectors in which there is no separate male or female component; each component can plug into any other similar component.

STANDARD PHONE JACKS

The standard phone jack is specified by a variety of different names, such as RJ and RG, which refer to their physical and electrical characteristics. These jacks consist of a male and a female component. The male component snaps into the female receptacle. The important point to note, however, is the number of conductors each type of jack can support.

Common configurations for phone jacks include support for four, six, or eight conductors. A typical example of a four-conductor jack, supporting two twisted-pairs, would be the one used for connecting most telephone handsets to their receivers.

A common six-conductor jack, supporting three twisted-pairs, is the RJ-11 jack used to connect most telephones to the telephone company or PBX systems. An example of an eight-conductor jack is the R-45 jack, which is intended for use under the ISDN system as the user-site interface for ISDN terminals.

For building wiring, the six-conductor and eight-conductor jacks are popular, with the eight-conductor jack increasing in popularity, as more corporations install twisted-pair in four-pair bundles for both voice and data. The eight-conductor jack, in addition to being used for ISDN, is also specified by several other popular applications, such as the new IEEE 802.3 10 BaseT standard for Ethernet over twisted-pair.

These types of jacks are often keyed, so that the wrong type of plug cannot be inserted into the jack. There are two kinds of keying – – side keying, and shift keying.

Side keying uses a piece of plastic that is extended to one side of the jack. This type is often used when multiple jacks are present.

Shift keying entails shifting the position of the snap connector to the left or right of the jack, rather than leaving it in its usual center position. Shift keying is more commonly used for data connectors than for voice connectors.

Note that while we say that these jacks are used for certain types of systems (data, voice, etc.) this is not any type of standard. They can be used as you please.

There are any number of pin type connectors available. The most familiar type is the RS-232 jack that is commonly used for computer ports. Another popular type of pin connector is the DB type connector, which is the round connector that is commonly used for computer keyboards.

The various types of pin connectors can be used for terminating as few as five (the DB type), or more than 50 (the RS type) conductors.

50-pin *champ* type connectors are often used with twisted-pair cables, when connecting to cross-connect equipment, patch panels, and communications equipment such as is used for networking.

CROSS CONNECTS

Cross connections are made at terminal *blocks*. A block is typically a rectangular, white plastic unit, with metal connection points. The most common type is called a punchdown block. This is the kind that you see on the back wall of a business, where the main telephone connections are made. The wire connections are made by pushing the insulated wires into their places. When "punched" down, the connector cuts through the insulation, and makes the appropriate connection.

Connections are made between punch-down blocks by using *patch cords*, which are short lengths of cable that can be terminated into the punch-down slots, or that are equipped with connectors on each end.

When different systems must be connected together, cross-connects are used.

CATEGORY CABLING

Category 1 cable is the old standard type of telephone cable, with four conductors colored green, red, black, and yellow. Also called quad cable.
Category 2 was an old IBM cabling system. It is almost never used for modern communications.
Category 3 cable is used for digital voice and data transmission rates up to 10 Mbps (Megabits per second). Common types of data transmission over this communications cable would be UTP Token Ring (4 Mbps) and 10Base-T (10 Mbps).
Category 4 cable is used for voice and data transmission rates up to 16 Mbps. A typical type of transmission over this system would be UTP Token Ring (16 Mbps).
Category 5 cable is used for sending voice and data at speeds up to 100Mbit/s (megabits per second).
This includes signals used under the FDDI communications standard.

INSTALLATION REQUIREMENTS

Article 800 of the NEC covers communication circuits, such as telephone systems and outside wiring for fire and burglar alarm systems. Generally these circuits must be separated from power circuits and grounded. In addition, all such circuits that run out of doors (even if only partially) must be provided with circuit protectors (surge or voltage supressors).

The requirements for these installations are as follows:

CONDUCTORS ENTERING BUILDINGS

If communications and power conductors are supported by the same pole, or run parallel in span, the following conditions must be met :

1. Wherever possible, communications conductors should be located below power conductors.
2. Communications conductors cannot be connected to crossarms .
3. Power service drops must be separated from communications service drops by at least 12 inches.

Above roofs, communications conductors must have the following clearances:

1. Flat roofs: 8 feet.
2. Garages and other auxiliary buildings: None required.
3. Overhangs, where no more than 4 feet of communications cable will run over the area: 18 inches.
4. Where the roof slope is 4 inches rise for every 12 inches horizontally: 3 feet.

Underground communications conductors must be separated from power conductors in manhole or handholes by brick, concrete, or tile partitions.

Communications conductors should be kept at least 6 feet away from lightning protection system conductors.

CIRCUIT PROTECTION

Protectors are surge arresters designed for the specific requirements of communications circuits. They are required for all aerial circuits not confined with a *block*. (Block here means city block.) They must be installed on all clrcults with a block that could accidentally contact power circuits over 300 volts to ground. They must also be listed for the type of installation.

Other requirements are the following:

Metal sheaths of any communications cables must be grounded or interrupted with an insulating joint as close as practicable to the point where they enter any building (such point of entrance being the place where the communications cable emerges through an exterior wall or concrete floor slab, or from a grounded rigid or intermediate metal conduit).

Grounding conductors for communications circuits must be copper or some other corrosion-resistant material, and have insulation suitable for the area in which it is installed.

Communications grounding conductors may be no smaller than No. 14.

The grounding conductor must be run as directly as possible to the grounding electrode, and be protected if necessary.

If the grounding conductor Is protected by metal raceway, it must be bonded to the grounding conductor on both ends.

CIRCUIT PROTECTION

Grounding electrodes for communications ground may be any of the following:
1. The grounding electrode of an electrical power system.
2. A grounded interior metal piping system (Avoid gas piping systems for obvious reasons.)
3. Metal power service raceway.
4. Power service equipment enclosures.
5. A separate grounding electrode.

If the building being served has no grounding electrode system, the following can be used as a grounding electrode:
1. Any acceptable power system grounding electrode. (See Section 250-81.)
2. A grounded metal structure.
3. A ground rod or pipe at least 5 feet long and 1/2 inch in diameter. This rod should be driven into damp (if possible) earth, and kept separate from any lightning protection system grounds or conductors.

Connections to grounding electrodes must be made with approved means .

If the power and communications systems use separate grounding electrodes, they must be bonded together with a No. 6 copper conductor. Other electrodes may be bonded also. This is not required for mobile homes.

For mobile homes, if there is no service equipment or disconnect within 30 feet of the mobile home wall, the communications circuit must have its own grounding electrode. In this case, or if the mobile home is connected with cord and plug, the communications circuit protector must be bonded to the mobile home frame or grounding terminal with a copper conductor no smaller than No. 12.

INTERIOR COMMUNICATIONS CONDUCTORS

Communications conductors must be kept at least 2 inches away from power or Class 1 conductors, unless they are permanently separated from them or unless the power or Class 1 conductors are enclosed in one of the following:
1. Raceway.
2. Type AC, MC, UF, NM, or NM cable, or metal-sheathed cable.

Communications cables are allowed in the same raceway, box, or cable with any of the following:
1. Class 2 and 3 remote-control, signaling, and power-limited circuits.
2. Power-limited fire protective signaling systems.
3. Conductive or nonconductive optical fiber cables.
4. Community antenna television and radio distribution systems.

Communications conductors are not allowed to be in the same raceway or fitting with power or Class 1 circuits.

Communications conductors are not allowed to be supported by raceways unless the raceway runs directly to the piece of equipment the communications circuit serves.

Openings through fire-resistant floors, walls, etc. must be sealed with an appropriate firestopping material.

Any communications cables used in plenums or environmental air-handling spaces must be listed for such use.

4-5

STANDARD TELECOM COLOR CODING

PAIR #	TIP (+) COLOR	RING (–) COLOR
1	White	Blue
2	White	Orange
3	White	Green
4	White	Brown
5	White	Slate
6	Red	Blue
7	Red	Orange
8	Red	Green
9	Red	Brown
10	Red	Slate
11	Black	Blue
12	Black	Orange
13	Black	Green
14	Black	Brown
15	Black	Slate
16	Yellow	Blue
17	Yellow	Orange
18	Yellow	Green
19	Yellow	Brown
20	Yellow	Slate
21	Violet	Blue
22	Violet	Orange
23	Violet	Green
24	Violet	Brown
25	Violet	Slate

25-PAIR COLOR CODING/ISDN
CONTACT ASSIGNMENTS

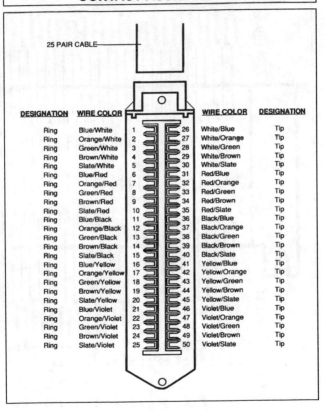

25 PAIR CABLE

DESIGNATION	WIRE COLOR			WIRE COLOR	DESIGNATION
Ring	Blue/White	1	26	White/Blue	Tip
Ring	Orange/White	2	27	White/Orange	Tip
Ring	Green/White	3	28	White/Green	Tip
Ring	Brown/White	4	29	White/Brown	Tip
Ring	Slate/White	5	30	White/Slate	Tip
Ring	Blue/Red	6	31	Red/Blue	Tip
Ring	Orange/Red	7	32	Red/Orange	Tip
Ring	Green/Red	8	33	Red/Green	Tip
Ring	Brown/Red	9	34	Red/Brown	Tip
Ring	Slate/Red	10	35	Red/Slate	Tip
Ring	Blue/Black	11	36	Black/Blue	Tip
Ring	Orange/Black	12	37	Black/Orange	Tip
Ring	Green/Black	13	38	Black/Green	Tip
Ring	Brown/Black	14	39	Black/Brown	Tip
Ring	Slate/Black	15	40	Black/Slate	Tip
Ring	Blue/Yellow	16	41	Yellow/Blue	Tip
Ring	Orange/Yellow	17	42	Yellow/Orange	Tip
Ring	Green/Yellow	18	43	Yellow/Green	Tip
Ring	Brown/Yellow	19	44	Yellow/Brown	Tip
Ring	Slate/Yellow	20	45	Yellow/Slate	Tip
Ring	Blue/Violet	21	46	Violet/Blue	Tip
Ring	Orange/Violet	22	47	Violet/Orange	Tip
Ring	Green/Violet	23	48	Violet/Green	Tip
Ring	Brown/Violet	24	49	Violet/Brown	Tip
Ring	Slate/Violet	25	50	Violet/Slate	Tip

66 BLOCK WIRING & CABLE COLOR CODING

PAIR CODE		SIDE #1		SIDE #2
Pair 1	Tip 26	White/Blue		White/Blue
	Ring 1	Blue/White		Blue/White
Pair 2	Tip 27	White/Orange		White/Orange
	Ring 2	Orange/White		Orange/White
Pair 3	Tip 28	White/Green		White/Green
	Ring 3	Green/White		Green/White
Pair 4	Tip 29	White/Brown		White/Brown
	Ring 4	Brown/White		Brown/White
Pair 5	Tip 30	White/Slate		White/Slate
	Ring 5	Slate/White		Slate/White
Pair 6	Tip 31	Red/Blue		Red/Blue
	Ring 6	Blue/Red		Blue/Red
Pair 7	Tip 32	Red/Orange		Red/Orange
	Ring 7	Orange/Red		Orange/Red
Pair 8	Tip 33	Red/Green		Red/Green
	Ring 8	Green/Red		Green/Red
Pair 9	Tip 34	Red/Brown		Red/Brown
	Ring 9	Brown/Red		Brown/Red
Pair 10	Tip 35	Red/Slate		Red/Slate
	Ring 10	Slate/Red		Slate/Red
Pair 11	Tip 36	Black/Blue		Black/Blue
	Ring 11	Blue/Black		Blue/Black
Pair 12	Tip 37	Black/Orange		Black/Orange
	Ring 12	Orange/Black		Orange/Black
Pair 13	Tip 38	Black/Green		Black/Green
	Ring 13	Green/Black		Green/Black

4-8

66 BLOCK WIRING & CABLE COLOR CODING

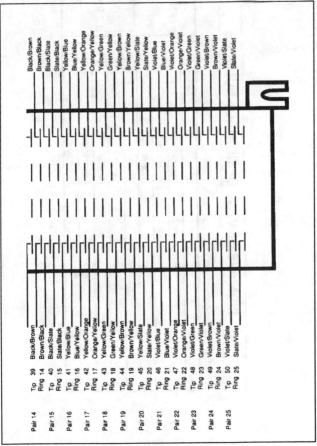

Pair 14	Tip 39	Black/Brown
	Ring 14	Brown/Black
Pair 15	Tip 40	Black/Slate
	Ring 15	Slate/Black
Pair 16	Tip 41	Yellow/Blue
	Ring 16	Blue/Yellow
Pair 17	Tip 42	Yellow/Orange
	Ring 17	Orange/Yellow
Pair 18	Tip 43	Yellow/Green
	Ring 18	Green/Yellow
Pair 19	Tip 44	Yellow/Brown
	Ring 19	Brown/Yellow
Pair 20	Tip 45	Yellow/Slate
	Ring 20	Slate/Yellow
Pair 21	Tip 46	Violet/Blue
	Ring 21	Blue/Violet
Pair 22	Tip 47	Violet/Orange
	Ring 22	Orange/Violet
Pair 23	Tip 48	Violet/Green
	Ring 23	Green/Violet
Pair 24	Tip 49	Violet/Brown
	Ring 24	Brown/Violet
Pair 25	Tip 50	Violet/Slate
	Ring 25	Slate/Violet

ISDN ASSIGNMENT OF CONTACT NUMBERS

Table-Contact assignments for plugs and jacks:

Contact Number	TE	NT	Polarity
1	Power source 3	Power sink 3	+
2	Power source 3	Power sink 3	-
3	Transmit	Receive	+
4	Receive	Transmit	+
5	Receive	Transmit	-
6	Transmit	Receive	-
7	Power sink 2	Power source 2	-
8	Power sink 2	Power source 2	+

8P8C (ISDN)

TYPICAL WIRING METHODS

LOOP SERIES WIRING

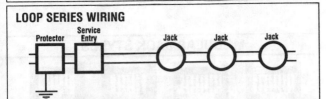

PARALLEL DISTRIBUTION WIRING

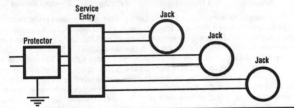

2-LINE SYSTEM

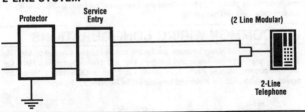

ELECTRONIC KEY SYSTEMS

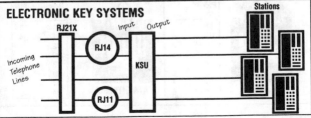

MODULAR JACK STYLES

8-Position

8-Position Keyed

6-Position

6-Position Modified

There are four basic modular jack styles. The 8-position and 8-position keyed modular jacks are commonly and incorrectly referred to as RJ45 and keyed RJ45 (respectively). The 6-position modular jack is commonly referred to as RJ11. Using these terms can sometimes lead to confusion since the RJ designations actually refer to very specific wiring configurations called Universal Service Ordering Codes (USOC). The designation 'RJ' means Registered Jack. Each of these 3 basic jack styles can be wired for different RJ configurations. For example, the 6-position jack can be wired as a RJ11C (1-Pair), RJ14C (2-Pair), or RJ25C (3-Pair) configuration. An 8-position jack can be wired for configurations such as RJ61C (4-Pair) and RJ48C. The keyed 8-position jack can be wired for RJ45S, RJ46S and RJ47S. The fourth modular jack style is a modified version of the 6-position jack (modified modular jack or MMJ). It was designed by DEC along with the modified modular plug (MMP) to eliminate the possibility of connecting DEC data equipment to voice lines and vice versa.

COMMON WIRING CONFIGURATIONS

The TIA and AT&T wiring schemes are the two that have been adopted by EIA/TIA-568. They are nearly identical except that pairs two and three are reversed. TIA is the preferred scheme because it is compatible with 1 or 2-pair USOC Systems. Either configuration can be used for Integrated Services Digital Network (ISDN) applications.

Pair ID	PIN #
T1	5
R1	4
T2	3
R2	6
T3	1
R3	2
T4	7
R4	8

TIA (T568A)

Pair ID	PIN #
T1	5
R1	4
T2	1
R2	2
T3	3
R3	6
T4	7
R4	8

AT&T (T568B)

COMMON WIRING CONFIGURATIONS

USOC wiring is available for 1-, 2-, 3-, or 4-pair systems. Pair 1 occupies the center conductors, pair 2 occupies the next two contacts out, etc. One advantage to this scheme is that a 6-position plug configured with 1, 2, or 3 pairs can be inserted into an 8-position jack and maintain pair continuity; a note of warning though, pins 1 and 8 on the jack may become damaged from this practice. A disadvantage is the poor transmission performance associated with this type of pair sequence.

Pair ID	PIN #
T1	5
R1	4
T2	3
R2	6
T3	2
R3	7
T4	1
R4	8

USOC 4-Pair

Pair ID	PIN #
T1	4
R1	3
T2	2
R2	5
T3	1
R3	6

USOC 1-, 2-, or 3-Pair

ETHERNET 10 BASE-T

Ethernet 10BASE-T wiring specifies an 8-position jack but uses only two pairs. These are pairs two and three of TIA schemes.

Pair ID	PIN #
T1	1
R1	2
T2	3
R2	6

COMMON WIRING CONFIGURATIONS

IBM Token-Ring

IBM Token-Ring wiring uses either an 8-position or 6-position jack. The 8-position format is compatible with TIA wiring schemes. The 6-position is compatible with 2 or 3-pair USOC wiring.

Pair ID	PIN #
T1	5
R1	4
T2	3
R2	6

DEC 3-Pair

DEC custom-designed wiring scheme is unique.

Pair ID	PIN #
T1	2
R1	3
T2	5
R2	4
T3	1
R3	6

MODULAR PLUG PAIR CONFIGURATIONS

It is important that the pairing of wires in the modular plug match the pairs in the modular jack as well as the horizontal and backbone wiring. If they don't, the data being transmitted may be paired with incompatible signals.

STRAIGHT THROUGH OR REVERSED?

Modular cords are used for two basic applications. One application uses them for patching between modular patch panels. When used in this manner modular cords should always be wired "straight-through" (pin 1 to pin 1, pin 2 to pin 2, pin 3 to pin 3, etc.). The second major application uses modular cords to connect the workstation equipment (PC, phone, FAX, etc.) to the modular outlet. These modular cords may either be wired "straight-through" or "reversed" (pin 1 to pin 6, pin 2 to pin 5, pin 3 to pin 4, etc.) depending on the system manufacturer's specifications. This "reversed" wiring is typical for most voice systems. The following is a guide to determine what type of modular cord you have.

HOW TO READ A MODULAR CORD

Align the plugs side-by-side with the contacts facing you and compare the wire colors from left to right. If the colors appear in the same order on both plugs, the cord is wired "straight-through" If the colors appear reversed on the second plug (from right to left) the cord is wired "reversed".

CHAPTER 5
FIBER OPTICS

TYPICAL OPTICAL BUDGETS

Wavelength [nm]	Type of Source	Fiber Core Diameter [μm]	Typical Optical Power Budget, [dB]	Spectral Width [nm]
850	LED	50	12	
	LED	6.25	16	
	LED	100	21	
	LED	any		30-50
	laser	all	30	
1300	LED	50	20	
	LED	62.5	24	
	LED	100	28	
	LED	any		60-190
1300	laser [multimode]	all	50	
1300	laser [singlemode]	9	27	0.5 - 5.0

FIBER OPTIC CONNECTORS
DATA COMMUNICATION STYLES

Style	Contact?	Keyed?	Pull Proof?	Wiggle-Proof?	Loss	Cost
ST®	Y	Y	N	N	0.3	6-10
906 SMA	N	N	Y	N	1.0	6-10
FDDI	Y	Y	Y	N	1.0	12-19
mini-BNC	N	N	N	N		10
905 SMA	N	N	Y	N	1.0	6-10
biconic	N	N	Y	N	1.0	8-25
ESCON	Y	Y	N	N	1.0	25
SC	Y	Y	Y	Y	0.3	7-14

TELEPHONE & HIGH PERFORMANCE STYLES

Style	Contact?	Keyed?	Pull Proof?	Wiggle-Proof?	Loss	Cost
ST®	Y	Y	N	N	0.3	6-10
biconic	N	N	Y	N	1.0	8-25
keyed biconic	N	Y	Y	N	1.0	8-25
FC/PC	Y	Y	Y	Y	0.3	4-15
FC	Y	Y	Y	Y	0.3	4-15
D4	Y	Y	Y	Y	0.3	12
SC	Y	Y	Y	Y	0.3	7-14

TYPICAL BANDWIDTH - DISTANCE PRODUCTS

Type of Fiber	Wavelength	Bandwidth - Distance Product
multimode step index POF	660 nm	5 MHz-km
multimode step index glass	850 nm	20 MHz-km
multimode graded index glass	850 nm	600 MHz-km
multimode graded index glass	1300 nm	1000-2500 MHz-km
singlemode glass	1310 nm	76,800-300,000 Mbps-km

OPTICAL CABLE JACKET MATERIALS & THEIR PROPERTIES

Jacket Materials	Properties
PVC	Affords normal mechanical protection. Usually specified for indoor use and general-purpose applications.
Hypalon	Has most of neoprene's properties, including ability to withstand extreme environments and flame retardancy. Has better thermal stability, and even greater oxidation and ozone resistance. Hypalon has superior resistance to radiation.
Polyethylene	Used in telephone cables. A tough, chemical- and moisture-resistant, relatively low-cost material. Since it burns, it is infrequently used in electronic applications.
Polyurethane	Has excellent abrasion resistance and low-temperature flexibility.
Thermoplastic Elastomer (TPE)	A less expensive jacketing material than neoprene or hypalon. Has many of the characteristics of rubber, along with excellent mechanical and chemical properties.
Nylon	Generally used over single conductors to improve their physical properties.

OPTICAL CABLE JACKET MATERIALS & THEIR PROPERTIES

Jacket Materials	Properties
Kynar (polyvinylidene fluoride)	A tough, abrasion- and cut- through-resistant, thermally stable and self-extinguishing material. It has low-smoke emission and is resistant to most chemicals. Its inherent stiffness limits its use as a jacket material. It has been approved for low-smoke applications.
Teflon FEP	Specified in fire alarm signal system cables. It will not emit smoke even when exposed to direct flame, is suitable for use at continuous temperatures of 200°C and is chemically inert.
Tefzel	Like Teflon FEP, it is a fluorocarbon and has many of its properties. Rated for 150°C, it is a tough, self-extinguishing material.
Irradiated Cross-Linked	Rated for 150°C operation. Cross-linking changes thermoplastic polyethylene to a thermosetting material with greater resistance to environmental stress cracking, cut-through, ozone, solvents and soldering than either low- or high-density polyethylene.
Zero Halogen Thermoplastic	A thermoplastic material with excellent flame retardancy properties. Does not emit toxic fumes when it burns. Originally designed for shipboard fiber applications, it can be used for any enclosed environment.

NEC CLASSIFICATIONS FOR OPTICAL CABLES

Application	Cable marking	UL test	Jacket type
General Purpose	OFN	UL-1581	PVC or
	OFC	UL-1581	Zero Halogen
Riser	OFNR	UL-1666	PVC or
	OFCR	UL01666	Zero Halogen
Plenum	OFNP	UL-910	RVC or
	OFCP	UL-910	Fluorocarbons

Cable Marking Explanation

OFN	Optical fiber, nonconductive (all dielectric)
OFC	Optical fiber, conductive (metal strength members)
OFNR	Optical fiber, nonconductive, riser
OFCR	Optical fiber, conductive, riser
OFNP	Optical fiber, nonconductive, plenum
OFCP	Optical fiber, conductive, plenum

COMMON FIBER TYPES							
Item Number	Diameter Core	Diameter Cladding	Index Profile	Primary Buffer Diameter	Attenuation (dB/km)	Bandwidth (MHz/km)	Numerical Aperture
1)	50µm	125µm	Graded	250µm	•3@0.85µm •1@1.3µm	200-800@0.85µm 200-800@1.3µm	0.20
2)	50µm	125µm	Graded	500µm	•4@0.85µm •2@1.3µm	200-800@0.85µm 200-800@1.3µm	0.20
3)	62.5µm	125µm	Graded	250µm	•3.5@0.85µm •1.5@1.3µm	100-300@0.85µm 100-800@1.3µm	0.275
4)	62.5µm	125µm	Graded	500µm	•3.5@0.85µm •1.5@1.3µm	100-300@0.85µm 100-800@1.3µm	0.275
5)	85µm	125µm	Graded	500µm	•4@0.85µm •2@1.3µm	100-200@0.85µm 200-400@1.3µm	0.26
6)	100µm	140µm	Graded	500µm	•5@0.85µm •3@1.3µm	100-300@0.85µm 100-500@1.3µm	0.30
7)	100µm	140µm	Step	500µm	•10@0.85µm	20@0.85µm	0.24
8)	100µm	140µm	Graded	160µm	•6@0.85µm	100@0.85µm	0.30
9)	200µm	230µm	Step(HCS)	500µm	•8@0.85µm	17@0.85µm	0.37
10)	200µm	240µm	Step	500µm	•10@0.85µm	20@0.85µm	0.24
11)	200µm	380µm	Step(PCS)	600µm	•10@0.85µm	8@0.85µm	0.40

Note items 7 and 9 are radiation-hard fibers. Item 8 is a high-temperature fiber.

COMPARISON OF LASER & LED LIGHT SOURCES

Characteristic	LED	Laser
Output power	Lower	Higher
Speed	Slower	Faster
Output pattern (NA)	Higher	Lower
Spectral width	Wider	Narrower
Single-mode compatibility	No	Yes
Ease of Use	Easier	Harder
Cost	Lower	Higher

COMPARISON OF BUFFER TYPES

Cable Parameter	Cable Structure	
	Loose Tube	Tight Buffer
Bend Radius	Larger	Smaller
Diameter	Larger	Smaller
Tensile Strength, Installation	Higher	Lower
Impact Resistance	Lower	Higher
Crush Resistance	Lower	Higher
Attenuation Change At Low Temperatures	Lower	Higher

INDICES OF REFRACTION

Material	Index	Light Velocity(km/s)
Vacuum	1.0	300,000
Air	1.0003	300,000
Water	1.33	225,000
Fused quartz	1.46	205,000
Glass	1.5	200,000
Diamond	2.0	150,000
Silicon	3.4	88,000
Gallium arsenide	3.6	83,000

TRANSMISSION RATES - DIGITAL TELEPHONE

Medium	Bit Rate (Mbps)	Voice Channels	Repeater Spacing (km)
Coaxial	1.5	24	1-2
	3.1	48	
	6.3	96	
	45	672	
	90	1,344	
Fiber	45	672	6-15 (multimode)
	90	1,344	30-40+ (single mode)
	180	2,688	
	405 to 435	6,048	
	565	8,064	
	1,700	24,192	

MISMATCHED FIBER CONNECTION LOSSES (Excess loss in db)

Receiving Fiber	Transmitting Fiber		
	62.5/125	85-125	100/140
50-125	0.9-1.6	3.0-4.6	4.7-9
62.5/125	—	0.9	2.1-4.1
85/125	—	—	0.9-1.4

POWER LEVELS OF FIBER OPTIC COMMUNICATION SYSTEMS

Network Type	Wavelength (nm)	Power Range (dBm)	Power Range (W)
Telecom	1300, 1550	+3 to -45	50nW to 2mW
Datacom	665, 790, 850, 1300	-10 to -30	1 100uW
CATV	1300, 1550	+10 to -6	250uW to 10mW

DETECTORS USED IN FIBER OPTIC POWER METERS

Detector Type	Wavelength Range (nm)	Power Range (dBm)	Comments
Silicon	400-1100	+10 to -70	
Germanium	800-1600	+10 to -60	-70 with small area detectors, +30 with attenuator windows
InGaAs	800-1600	+10 to -70	Small area detectors may overload at high power (>.0 dBm)

FIBER OPTIC TESTING REQUIREMENTS

Test Parameter	Instrument
Optical power (source output, receiver signal level)	Fiber optic power meter
Attenuation or loss of fibers, cables, and connectors	FO power meter and source, test kit or OLTS (optical loss test set)
Source wavelength*	FO spectrum analyzer
Backscatter (loss, length, fault location)	Optical time domain reflectometer (OTDR)
Fault location	OTDR, visual cable fault locator
Bandwidth/dispersion* (modal and chromatic)	Bandwidth tester or simulation software

* Rarely tested in the field

TYPICAL CABLE SYSTEM FAULTS

Fault	Cause	Equipment	Remedy
Bad connector	Dirt or damage	Microscope	Cleaning/ polishing retermination
Bad pigtail	Pigtail kinked	Visual fault locator	Straighten kink
Localized cable attenuation	Kinked cable	OTDR	Straighten kink
Distributed increase in cable attenuation	Defective cable or installation specifications exceeded	OTDR	Reduce stress/ replace
Lossy splice	Increase in splice Loss due to fiber stress in closure	OTDR Visual fault locator	Open and redress
Fiber break	Cable damage	OTDR Visual fault locator	Repair/replace

MOST COMMON CAUSES OF FAILURES IN FIBER OPTIC LANs

1	Broken fibers at connector joints
2	Broken fibers at patch panels
3	Cables damaged at patch panels
4	Fibers broken at patch panels
5	Cables cut in ceilings and walls
6	Cables cut through outside construction
7	Contaminated connections
8	Broken jumpers
9	Too much loss
10	Too little loss (overdriving the receiver)
11	Improper cable rolls
12	Miskeyed connectors
13	Transmission equipment failure
14	Power failure

FIBER OPTIC LABOR UNITS

Labor Item	Labor Units (Hours)	
	Normal	Difficult
Optical fiber cables, per foot:		
1-4 fibers, in conduit	0.016	0.02
1-4 fibers, accessible locations	0.014	0.018
12-24 fibers, in conduit	0.02	0.025
12-24 fibers, accessible locations	0.018	0.023
48 fibers, in conduit	0.03	0.038
48 fibers, accessible locations	0.025	0.031
72 fibers, in conduit	0.04	0.05
72 fibers, accessible locations	0.032	0.04
144 fibers, in conduit	0.05	0.065
144 fibers, accessible locations	0.04	0.05
Hybrid cables:		
1-4 fibers, in conduit	0.02	0.025
1-4 fibers, accessible locations	0.017	0.021
12-24 fibers, in conduit	0.024	0.03
12-24 fibers, accessible locations	0.022	0.028
Testing, per fiber	0.12	0.24
Splices, including prep and failures, trained workers:		
fusion	0.30	0.45
mechanical	0.40	0.50
array splice, 12 fibers	1.00	1.30
Coupler (connector-connector)	0.15	0.25
Terminations, including prep and failures, trained workers:		
polishing required	0.40	0.60
no-polish connectors	0.30	0.45
FDDI dual connector, including terminations	0.80	1.00
Miscellaneous:		
cross-connect box, 144 fibers, not including splices	3.00	4.00
splice cabinet	2.00	2.50
splice case	1.80	2.25
breakout kit, 6 fibers	1.00	1.40
tie-wraps	0.01	0.02
wire markers	0.01	0.01

Fiber Optic Safety Rules

1. Keep all food and beverages out of the work area. If fiber particles are ingested, they can cause internal hemorrhaging.
2. Wear disposable aprons to minimize fiber particles on your clothing. Fiber particles on your clothing can later get into food, drinks, and/or be ingested by other means.
3. Always wear protective gloves and safety glasses with side shields. Treat fiber optic splinters the same as you would glass splinters.
4. Never look directly into the end of fiber cables until you are positive that there is no light source at the other end. Use a fiber optic power meter to make certain the fiber is dark. When using an optical tracer or continuity checker, look at the fiber from an angle at least six inches away from your eye to determine if the visible light is present.
5. Only work in well ventilated areas.
6. Contact wearers must not handle their lenses until they have thoroughly washed their hands.
7. Do not touch your eyes while working with fiber optic systems until your hands have been thoroughly washed.

COMMON FIBER OPTIC CONNECTORS

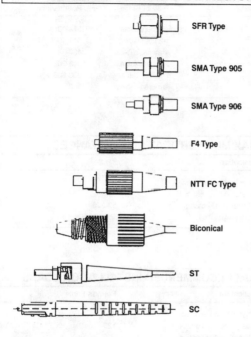

SFR Type

SMA Type 905

SMA Type 906

F4 Type

NTT FC Type

Biconical

ST

SC

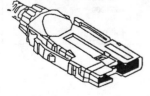

FDDI Fixed Shroud Duplex Connector

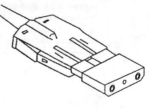

ESCON Connector

OPTICAL CABLE CRUSH STRENGTHS

Characteristic	Type of Cable	Pounds/Inch
Long-term crush load	>6 fibers/cable	57-400
	1-2 fiber cables	314-400
	Armored cables	450
Short-term crush load	>6 fibers/cable	343-900
	1-2 fiber cables	300-800
	Armored cables	600

MAXIMUM VERTICAL RISE DISTANCES

Application	Feet
1 fiber in raceway or tray	90
2 fiber in duct or conduit	50-90
Multifiber (6-12) cables	50-375
Heavy duty cables	1000-1640a

MAXIMUM RECOMMENDED INSTALLATION LOADS

Application	Pounds Force
1 fiber in raceway or tray	67
1 fiber in duct or conduit	125
2 fiber in duct or conduit	200
Multifiber (6-12) cables	250-500
Direct burial cables	600-800
Lashed aerial cables	>300
Self-support aerial cables	>600

CABLE SELECTION CRITERIA

1. Current and future bandwidth requirements
2. Acceptable attenuation rate
3. Length of cable
4. Cost of installation
5. Mechanical requirements (ruggedness, flexibility, flame retardance, low smoke, cut-through resistance)
6. UL/NEC requirements
7. Signal source (coupling efficiency, power output, receiver sensitivity)
8. Connectors and terminations
9. Cable dimension requirements
10. Physical environment (temperature, moisture, location)
11. Compatibility with any existing systems

FIBER OPTIC DATA NETWORK STANDARDS

Network	IEEE802.3 FOIRL	IEEE802.3 10baseF	IEEE802.5 Token Ring	ANSI X3T9.5 FDDI	ESCON IBM
Bitrate (MB/s)	10	10	4/16	100	200
Architecture	Link	Star	Ring	Ring	Branch
Fiber type	MM, 62.5	MM, 62.5	MM, 62.5	MM/SM	MM/SM
Link length (km)	2	—	—	2/60	3/20
Wavelength (nm)	850	850	850	1300	1300
Margin (dB, MM/SM)	8	—	12	11/27	8*(11)/16
Fiber FW (mHz-km)	150	150	150	500	500
Connector	SMA	ST	FDDI	FDDI	ESCON

FIBER TYPES AND SPECIFICATIONS

Fiber Type	Core/Cladding Diameter (m)	Attenuation Coefficient (dBkm) 850 nm	1300 nm	1550 nm	Bandwidth (MHz-km)
Multimode/Plastic	1mm	(1 dB/m	@665 nm)		Low
Multimode/Step Index	200/240	6			50
Multimode/Graded Index	50/125	3	1		600
	62.5/125	3	1		500
	85/125	3	1		500
	100/140	3	1		300
Single mode	8-9/125		0.5	0.3	high

NOTES

CHAPTER 6
MOTORS

DESIGNING MOTOR CIRCUITS

For one motor:
1. Determine full-load current of motor(s).
2. Multiply full-load current x 1.25 to determine minimum conductor ampacity.
3. Determine wire size.
4. Determine conduit size.
5. Determine minimum fuse or circuit breaker size.
6. Determine overload rating.

For more than one motor:
1. Perform steps 1 through 6 as shown above for each motor.
2. Add full-load current of all motors, plus 25% of the full-load current of the largest motor to determine minimum conductor ampacity.
3. Determine wire size.
4. Determine conduit size.
5. Add the fuse or circuit breaker size of the largest motor, plus the full-load currents of all other motors to determine the maximum fuse or circuit breaker size for the feeder.

AVERAGE ELECTRIC MOTOR SPECS

NOTE: Use the following table as a general guide only! These numbers are for normal fan, furnace, appliance, pump, and normal duty applications. The exact specifications for any given motor can vary greatly from those listed below. For 230V motors simply divide the indicated amps by 2.

Specs for 115 volt, 60 Hz, 1 Phase, AC Electric Motors

Motor Horsepower	RPM	Full Load Amps
1/20	1550	2.5
1/15	1550	2.8
1/12	1725	2.2-2.8
	1550	4.1
	850	3.2
1/10	1550	3.5
	1050	3.4-4.2
1/8	1725	1.8-2.7
	1140	3.8
	1075	1.8-5.0
1/6	1725	3.3-4.7
	1550	4.0-4.8
	1140	4.0-4.9
	1075	2.4-5.0
1/4	1725	4.4-6.3
	1625	3.1-3.6
	1140	5.6-6.8
	1075	3.4-6.8
	850	6.9
1/3	3450	5.6-6.5
	1725	5.3-6.8
	1140	5.0-7.2
	1075	5.1
1/2	3450	9.8
	1725	7.0-9.2
	1075	7.3
3/4	3450	11.8
	1725	11.6
	1075	9.5
1	3450	13.0-15.0
	1725	13.6-16.0
1-1/2	3450	16.4-19.6
	1725	19.6
2	3450	19-23

The above general specifications are based on motor data from the 1988 Graingers Catalog, Chicago, Illinois.

NEMA ELECTRIC MOTOR FRAMES

Motor Frame	NEMA Frame Dimension - Inches						
	D	E	F	U	V	M+N	Keyway
42	2-5/8	1-3/4	27/32	3/8		4-1/32	
48	2	2-1/8	1-3/8	1/2		5-3/8	
56	3-1/2	2-7/16	1-1/2	5/8		6-1/8	3/16x3/32
66	4-1/8	2-15/16	2-1/2	3/4		7-7/8	3/16x3/32
143T	3-1/2	2-3/4	2	7/8	2	6-1/2	3/16x3/32
145T	3-1/2	2-3/4	2-1/2	7/8	2	7	3/16x3/32
182	4-1/2	3-3/4	2-1/4	7/8	2	7-1/4	3/16x3/32
182T	4-1/2	3-3/4	2-1/4	1-1/8	2-1/2	7-3/4	1/4x1/8
184	4-1/2	3-3/4	2-3/4	7/8	2	7-3/4	3/16x3/32
184T	4-1/2	3-3/4	2-3/4	1-1/8	2-1/2	8-1/4	1/4x1/8
213	5-1/4	4-1/4	2-3/4	1-1/8	2-3/4	9-1/4	1/4x1/8
213T	5-1/4	4-1/4	2-3/4	1-3/8	3-1/8	9-5/8	5/16x5/32
215	5-1/4	4-1/4	3-1/2	1-1/8	2-3/4	10	1/4x1/8
215T	5-1/4	4-1/4	3-1/2	1-3/8	3-1/8	10-3/8	5/16x5/32
254T	6-1/4	5	4-1/8	1-5/8	3-3/4	12-3/8	3/8x3/16
254U	6-1/4	5	4-1/8	1-3/8	3-1/2	12-1/8	5/16x5/32
256T	6-1/4	5	5	1-5/8	3-3/4	13-1/4	3/8x3/16
256U	6-1/4	5	5	1-3/8	3-1/2	13	5/16x5/32
284T	7	5-1/2	4-3/4	1-7/8	4-3/8	14-1/8	1/2x1/4
284TS	7	5-1/2	4-3/4	1-5/8	3	12-3/4	3/8x3/16
284U	7	5-1/2	4-3/4	1-5/8	4-5/8	14-3/8	3/8x3/16
286T	7	5-1/2	5-1/2	1-7/8	4-3/8	14-7/8	1/2x1/4
286U	7	5-1/2	5-1/2	1-5/8	4-5/8	15-1/8	3/8x3/16
324T	8	6-1/4	5-1/4	2-1/8	5	15-3/4	1/2x1/4
324U	8	6-1/4	5-1/4	1-7/8	5-3/8	16-1/8	1/2x1/4
326T	8	6-1/4	6	2-1/8	5	16-1/2	1/2x1/4
326TS	8	6-1/4	6	1-7/8	3-1/2	15	1/2x1/4
326U	8	6-1/4	6	1-7/8	5-3/8	16-7/8	1/2x1/4
364T	9	7	5-5/8	2-3/8	5-5/8	17-3/8	5/8x5/16
364U	9	7	5-5/8	2-1/8	6-1/8	17-7/8	1/2x1/4
365T	9	7	6-1/8	2-3/8	5-5/8	17-7/8	5/8x5/16
365U	9	7	6-1/8	2-1/8	6-1/8	18-3/8	1/2x1/4
404T	10	8	6-1/8	2-7/8	7	20	3/4x3/8
404U	10	8	6-1/8	2-3/8	6-7/8	19-7/8	5/8x5/16
405T	10	8	6-7/8	2-7/8	7	20-3/4	3/4x3/8
405U	10	8	6-7/8	2-3/8	6-7/8	20-5/8	5/8x5/16
444T	11	9	7-1/4	3-3/8	8-1/4	23-1/4	7/8x7/16
444U	11	9	7-1/4	2-7/8	8-3/8	23-3/8	3/4x3/8
445T	11	9	8-1/4	3-3/8	8-1/4	24-1/4	7/8x7/16
445U	11	9	8-1/4	2-7/8	8-3/8	24-3/8	3/4x3/8

The above standards were established by the National Electrical Manufacturers Association (NEMA)

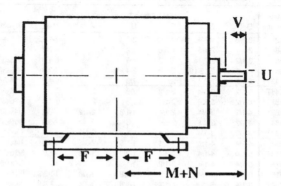

Frame dimensions for previous page.

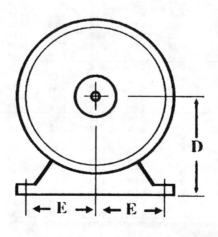

3 PHASE ELECTRIC MOTOR SPECS

HP	Full Load Amps 230V(460V)	Wire Size Minimum (AWG-Rubber) 230V(460V)	Conduit Size Inches 230V(460V)
1	3.3(1.7)	14(14)	1/2(1/2)
1.5	4.7(2.4)	14(14)	1/2(1/2)
2	6(3.0)	14(14)	1/2(1/2)
3	9(4.5)	14(14)	1/2(1/2)
5	15(7.5)	12(14)	1/2(1/2)
7.5	22(11)	8(14)	3/4(1/2)
10	27(14)	8(12)	3/4(1/2)
15	38(19)	6(10)	1-1/4(3/4)
20	52(26)	4(8)	1-1/4(3/4)
25	64(32)	3(6)	1-1/4(1-1/4)
30	77(39)	1(6)	1-1/2(1-1/4)
40	101(51)	00(4)	2(1-1/4)
50	125(63)	000(3)	2(1-1/4)
60	149(75)	200M(1)	2-1/2(1-1/2)
75	180(90)	0000(0)	2-1/2(2)
100	245(123)	500M(000)	3(2)
125	310(155)	750M(0000)	3-1/2(2-1/2)
150	360(180)	1000M(3000M)	4(2-1/2)
200	480(240)	NR(500M)	NR(3)
250	580(290)	NR(NR)	NR(NR)
300	696(348)	NR(NR)	NR(NR)

NR indicates "Not Recommended" and "M" indicates M.C.M (1000 Circular Mils).

Note that starting currents for the above motors can be many times the Full Load Amps and fuses must be adjusted accordingly. If the powerline becomes too long, voltage drop will exceed safe limits and the wire size should be adjusted to the next larger (smaller AWG number) gauge wire. See the Copper Wire Specifications table in this chapter for more specific information on wire.

The above specifications are from the *National Electrical Code*.

DC MOTOR WIRING SPECS

HP	Full Load Amps 115V(230V)	Wire Size Minimum (AWG-Rubber) 115V(230V)	Conduit Size Inches 115V(230V)
1	8.4(4.2)	14(14)	1/2(1/2)
1.5	12.5(6.3)	12(14)	1/2(1/2)
2	16.1(8.3)	10(14)	3/4(1/2)
3	23(12.3)	8(12)	3/4(1/2)
5	40(19.8)	6(10)	1(3/4)
7.5	58(28.7)	3(6)	1-1/4(1)
10	75(38)	1(6)	1-1/2(1)
15	112(56)	00(4)	2(1-1/4)
20	140(74)	000(1)	2(1-1/2)
25	184(92)	300M(0)	2-1/2(2)
30	220(110)	400M(00)	3(2)
40	292(146)	700M(0000)	3-1/2(2-1/2)
50	360(180)	1000M(300M)	4(2-1/2)
60	NR(215)	NR(400M)	NR(3)
75	NR(268)	NR(600M)	NR(3-1/2)
100	NR(355)	NR(1000M)	NR(4)

NR indicates "Not Recommended" and "M" indicates M.C.M (1000 Circular Mils).
The above specifications are based on data from the *National Electrical Code*.

HP vs TORQUE vs RPM – MOTORS

Torque in inch Pounds-force @ Motor R.P.M.

HP	3450	2000	1725	1550	1140	1050
1	18	32	37	41	55	60
1.5	27	47	55	61	83	90
2	37	63	73	81	111	120
3	55	95	110	122	166	180
5	91	158	183	203	276	300
7.5	137	236	274	305	415	450
10	183	315	365	407	553	600
15	274	473	548	610	829	900
20	365	630	731	813	1106	1200
25	457	788	913	1017	1382	1501
30	548	945	1096	1220	1659	1801
40	731	1261	1461	1626	2211	2401
50	913	1576	1827	2033	2764	3001
60	1096	1891	2192	2440	3317	3601
70	1279	2206	2558	2846	3870	4202
80	1461	2521	2923	3253	4423	4802
90	1644	2836	3288	3660	4976	5402
100	1827	3151	3654	4066	5529	6002
125	2284	3939	4567	5083	6911	7503
150	2740	4727	5480	6099	8293	9004
175	3197	5515	6394	7116	9675	10504
200	3654	6303	7307	8132	11057	12005
225	4110	7090	8221	9149	12439	13505
250	4567	7878	9134	10165	13821	15006
275	5024	8666	10047	11182	15203	16507
300	5480	9454	10961	12198	16586	18007
350	6394	11029	12788	14231	19350	21008
400	7307	12605	14614	16265	22114	24010
450	8221	14181	16441	18298	24878	27011
500	9134	15756	18268	20331	27643	30012
550	10047	17332	20095	22364	30407	33013
600	10961	18908	21922	24397	33171	36014

$$\text{Torque in inch Pounds-force} = \frac{\text{Horsepower} \times 63025}{\text{Motor RPM}}$$

To convert to Foot Pounds-force, divide the torque by 12.

HP vs TORQUE vs RPM – MOTORS

Torque in inch Pounds-force @ Motor R.P.M.

HP	1000	850	750	600	500	230
1	63	74	84	105	126	274
1.5	95	111	126	158	189	411
2	126	148	168	210	252	548
3	189	222	252	315	378	822
5	315	371	420	525	630	1370
7.5	473	556	630	788	945	2055
10	630	741	840	1050	1261	2740
15	945	1112	1261	1576	1891	4110
20	1261	1483	1681	2101	2521	5480
25	1576	1854	2101	2626	3151	6851
30	1891	2224	2521	3151	3782	8221
40	2521	2966	3361	4202	5042	10961
50	3151	3707	4202	5252	6303	13701
60	3782	4449	5042	6303	7563	16441
70	4412	5190	5882	7353	8824	19182
80	5042	5932	6723	8403	10084	21922
90	5672	6673	7563	9454	11345	24662
100	6303	7415	8403	10504	12605	27402
125	7878	9268	10504	13130	15756	34253
150	9454	11122	12605	15756	18908	41103
175	11029	12976	14706	18382	22059	47954
200	12605	14829	16807	21008	25210	54804
225	14181	16683	18908	23634	28361	61655
250	15756	18537	21008	26260	31513	68505
275	17332	20390	23109	28886	34664	75356
300	18908	22244	25210	31513	37815	82207
350	22059	25951	29412	36765	44118	95908
400	25210	29659	33613	42017	50420	109609
450	28361	33366	37815	47269	56723	123310
500	31513	37074	42017	52521	63025	137011
550	34664	40781	46218	57773	69328	150712
600	37815	44488	50420	63025	75630	164413

$$\text{Torque in inch Pounds-force} = \frac{\text{Horsepower} \times 63025}{\text{Motor RPM}}$$

NOTE: Ratings below 500 RPM are for gear motors.
To convert to Foot Pounds-force, divide the torque by 12.

HP vs TORQUE vs RPM – MOTORS

Torque in inch Pounds-force @ Motor R.P.M.

HP	190	155	125	100	84	68
1	332	407	504	630	750	927
1.5	498	610	756	945	1125	1390
2	663	813	1008	1261	1501	1854
3	995	1220	1513	1891	2251	2781
5	1659	2033	2521	3151	3751	4634
7.5	2488	3050	3782	4727	5627	6951
10	3317	4066	5042	6303	7503	9268
15	4976	6099	7563	9454	11254	13903
20	6634	8132	10084	12605	15006	18537
25	8293	10165	12605	15756	18757	23171
30	9951	12198	15126	18908	22509	27805
40	13268	16265	20168	25210	30012	37074
50	16586	20331	25210	31513	37515	46342
60	19903	24397	30252	37815	45018	55610
70	23220	28463	35294	44118	52521	64879
80	26537	32529	40336	50420	60024	74147
90	29854	36595	45378	56723	67527	83415
100	33171	40661	50420	63025	75030	92684
125	41464	50827	63025	78781	93787	115855
150	49757	60992	75630	94538	112545	139026
175	58049	71157	88235	110294	131302	162197
200	66342	81323	100840	126050	150060	185368
225	74635	91488	113445	141806	168817	208539
250	82928	101653	126050	157563	187574	231710
275	91220	111819	138655	173319	206332	254881
300	99513	121984	151260	189075	225089	278051
350	116099	142315	176470	220588	262604	324393
400	132684	162645	201680	252100	300119	370735
450	149270	182976	226890	283613	337634	417077
500	165855	203306	252100	315125	375149	463419
550	182441	223637	277310	346638	412664	509761
600	199026	243968	302520	378150	450179	556103

Torque in inch Pounds-force = $\dfrac{\text{Horsepower} \times 63025}{\text{Motor RPM}}$

NOTE: Ratings below 500 RPM are for gear motors.
To convert to Foot Pounds-force, divide the torque by 12.

TIPS ON SELECTING MOTORS

First step in selecting a motor for a particular drive is to obtain data listed below. Motors generally operate at best power factor and efficiency when fully loaded.

TORQUE Starting torque needed by load must be less than required starting torque of proposed motor. Motor torque must never fall below driven machine's torque needs in going from standstill to full speed.

Torque requirements of some loads may fluctuate between wide limits. Although average torque may be low, many torque peaks may be well above full-load torque. If load torque impulses are repeated frequently (air compressor), it's best to use a high-slip motor with a flywheel. But if load is generally steady at full load, you can use a more efficient low-slip motor. Only in this case any intermittent load peaks are taken directly by the motor and reflect back into power system. Also breakdown (maximum motor torque) must be higher than load-peak torque.

ENCLOSURES Atmospheric conditions surrounding motor determine enclosure used. The more enclosed a motor, the more it costs and the hotter it tends to run. Totally-enclosed motors may require a larger frame size for a given hp than open or protected motors.

INSULATION This, likewise, is determined by surrounding atmosphere and operating temperature. Ambient (room) temperature is generally assumed to be 40°C. Total temperature rise reaches directly influences insulation life. Each 10°C rise in max temp halves effective life of *Class A* and *B* insulations.

Motor temperature rise is maximum temperature (over ambient) measured with an external thermometer. "Hot-spot" allowance takes care of temperature difference between external reading and hottest spot within windings. Service-factor allows for continuous overload - 15% for general-purpose open, protected or drip-proof motors.

VARIABLE CYCLE Where load varies according to some regular cycle, it would not be economical to select a motor that matches the peak load.

Instead (for ac induction motors where speed is not varied) calculate hp needed on the *root-means-square* (rms) basis, example right. *Rms* hp is equivalent continous hp that would produce same heat in motor as cycle operation. Torque-speed relation of motor should still match that of load.

FACTS TO CONSIDER

REQUIREMENTS OF DRIVEN MACHINE:
1. Hp needed
2. Torque range
3. Operating cycle-frequency of starts and stops
4. Speed
5. Operating position-horizontal, vertical or tilted
6. Direction of rotation
7. Endplay and thrust
8. Ambient (room) temperature
9. Surrounding conditions-water, gas, corrosion, dust, outdoor, etc.

ELECTRICAL SUPPLY:
1. Voltage of power system
2. Number of phases
3. Frequency
4. Limitations on starting current
5. Effect of demand, energy on power rates

TYPES OF ENCLOSURES

OPEN-TYPE
has full openings in frame and endbells for maximum ventilation, is lowest cost enclosure.

SEMI-PROTECTED
has screens in top openings to keep out falling objects. PROTECTED has screens in bottom too.

DRIP-PROOF
has upper parts covered to keep out drippings falling at angle not over 15° from vertical

SPLASH-PROOF
is baffled at bottom to keep out particles coming at angle not over 100° from vertical

TOTALLY-ENCLOSED
can be non-ventilated, separately ventilated, or explosion proof for hazzardous atmospheres

FAN-COOLED
totally-enclosed motor has double covers. Fan, behind vented outer shroud, is run by motor

INSULATION, TEMPERATURE

CLASS A (cotton, silk, paper or other organics impregnated with insulating varnish) is considered standard for most applications, allows 105° C total temperature
40°C ambient
40°C rise by thermometer
15°C "hot-spot" allowance
10°C service factor
105°C total temperature
CLASS B (mica, asbestos, fiber-glass, other inorganics) allows 130°C total temperature
40°C ambient
70°C rise by thermometer
20°C "hot-spot" allowance
130°C total temperature

CLASS H (including silicone family) is for special high-temperature applications
SPECIAL CLASS A is highly resistant, but not "proof," against severe moisture, dampness; conductive, corrosive or abrasive dusts and vapors
TROPICAL is for excessive moisture, high ambients, corrosion, fungus, vermin, insects

SQUIRREL-CAGE MOTORS

Squirrel-cage induction motors are classed by National Electrical Manufacturers Assn. (NEMA) according to locked-rotor torque, breakdown torque, slip, starting current, etc. Common types are Class B, C and D.
CLASS B most common type, has normal starting torque, low starting current. Locked-rotor torque (minimum torque at standstill and full voltage) is not less than 100% full-load for 2 and 4-pole motors, 200 hp and less; 40 to 75% for larger 2 pole motors; 50 to 125% for larger 4-pole motors.
CLASS C features high starting torque (locked-rotor over 200%), low starting current. Breakdown torque not less than 190% full-load torque. Slip at full load is between 1 1/2 and 3%.
CLASS D have high slip, high starting torque, low starting current; are used on loads with high intermittent peaks. Driven machine usually has high-inertia flywheel. At no load motor has little slip; when peak load is applied, motor slip increases. Speed reduction lets driven machine absorb energy from flywheel rather than power line.
STARTING Full-voltage, across -the - line starting is used where power supply permits and full-voltage torque and acceleration are not objectionable. Reduction in starting kva cuts locked-rotor and accelerating torques.
APPLICATIONS of squirrel-cage and other motors are listed in table, pages 26 and 27.

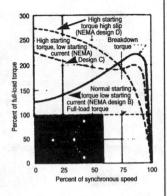

WOUND-ROTOR MOTORS

The wound-rotor (slip-ring) induction motor's rotor winding connects through slip-rings to an external resistance that is cut in and out by a controller.

RESISTANCE vs TORQUE Resistance of rotor winding affects torque developed at any speed. A high-resistance rotor gives high starting torque with low starting current. But low slip at full load, good efficiency and moderate rotor heating takes a low-resistance rotor. Left-hand curves show rotor-resistance effect on torque. With all resistance in, R_1, full-load starting torque is developed in less than 150% full-load current. Successively shorting out steps, at standstill, develops about 225% full-load torque at R_4. Cutting out more reduces standstill torque. Motor operates like a squirrel-cage motor when all resistance is shorted out.

SPEED CONTROL. Having resistance left in, decreases speed regulation. Righthand curves for a typical motor show that with only two steps shorted out, motor operates at 65% synchronous speed because motor torque equals load torque at that speed. But if load torque drops to 50%, motor shoots forward to about 65% synchronous speed.

However slip rings are normally shorted after motor comes up to speed. Or, for short-time peak loads, motor is operated with a step or two of resistance cut in. At light loads, motor runs near synchronous speed. When peak loads come on, speed drops; flywheel effect of motor and load cushions power supply from load peak.

OTHER FEATURES In addition to high-starting-torque, low-starting-current applications, wound-rotor motors are used (1) for high-inertia loads where high slip losses that would have to be dissipated in the rotor of a squirrel-cage motor, in coming up to speed, can be given off as heat in wound-rotor's external resistance (2) where frequent starting, stopping and speed control are needed (3) for continuous operation at reduced speed. (Example: boiler draft fan - combustion control varies external resistance to adjust speed, damper regulates air flow between step speeds and below 50% speed.)

CONTROLS used are across-the-line starters with proper protection (fuses, breakers, etc.) arid secondary control with 5 to 7 resistance steps.

SYNCHRONOUS MOTORS

Synchronous motors run at a fixed or synchronous speed determined by line frequency and number of poles in machine. (Rpm=120xfrequency/number or poles.) Speed is kept constant by locking action of an externally excited dc field. Efficiency is 1 to 3% higher than that of same-size-and-speed induction or dc motors. Also synchronous motors can be operated at power factors from 1.0 down to 0.2 leading for plant power-factor correction. Standard ratings are 1.0 and 0.9 leading pf; machines rated down near 0.2 leading are called synchronous condensers.

STARTING Pure synchronous motors are not self-starting; so in practice they are built with damper or amortisseur windings. With the field coil shorted through discharge resistor, damper winding acts like a squirrel-cage rotor to bring motor practically to synchronous speed; then field is applied and motor pulls into synchronism, providing motor has developed sufficient "pull-in" torque. Once in synchronism, motor keeps constant speed as long as load torque does not exceed maximum or "pull-out" torque; then machine drops out of synchronism. Driven machine is usually started without load. Low-speed motors may be direct connected.

FIELD AND PF While motors are rated for specific power factors...at constant power, increasing dc field current causes power factor to lead, decreasing field tends to make pf lag. But either case increases copper losses.

TYPES Polyphase synchronous motors in general use are: (1) high-speed motors, 500 rpm up, (a) general-purpose, 500 to 1800 rpm, 200 hp and below, (b) high-speed units over 200, hp including most 2-pole motors - (2) low-speed motors below 500 rpm, (3) special high-torque motors.

COSTS Small high-speed synchronous motors cost more than comparable induction motors, but with low-speed and large high-speed motor, costs favor the synchronous motor. Cost of leading-pf motors increases approximately inversely proportional to the decrease from unity power factor.

DC MOTORS

Chief reason for using dc motors, assuming normal power source is ac, lies in the wide and economical ranges possible of speed control and starting torques. But for constant-speed service, ac motors are generally preferred because they are more rugged and have lower first cost.

STARTING TORQUE With a shunt motor, torque is proportional to armature current, because field flux remains practically constant for a given setting of the field rheostat. However, the flux of a series field is affected by the current through it. At light loads, flux varies directly with a current, so torque varies as the square of the current. The compound motor (usually cumulative) lies in between the shunt and series motors as to torque.

Upper limit of current input on starting is usually 1.5 to 2 times full-load current to avoid over-heating the commutator, excessive fedder drops or peaking generator. Shunt-motor starting boxes usually allow 125% current at first notch. So motor can develop 125% starting torque. Series

motors can develop higher starting torques at same current, since torque increases as current squared. Compound motors develop starting torques higher than shunt motors according to amount of compounding.

SPEED CONTROL Shunt motor speeds drop only slightly (5% or less) from no load to full load. Decreasing field current raises speed; increasing field reduces speed. But speed is still practically constant for any one field setting. Speed can be controlled by resistance in the armature circuit but regulation is poor.

Series motor speeds decrease much more with increased load, and, conversely, begin to race at low loads, dangerously so if load is completely removed. Speed can be reduced by adding resistance into the armature circuit, increased by shunting the series filed with resistance or short-circuiting series turns.

Compound motors have less constant speed than shunt motors and can be controlled by shunt-field rheostat.

SUMMARY OF APPLICATIONS

DC and Single phase motors

Speed Regulation	Speed Control	Starting Torque	Pull-out Torque	Applications
SERIES				
Varies inversely as the load. Races on light loads and full voltage	Zero to maximum depending on control and load	High. Varies as square of the voltage. Limited by commutation, heating and line capacity	High. Limited by commutation, heating and line capacity	Where high starting torque is required and speed can be regulated. Traction, bridges, hoists, gates, car dumpers, car retarders, etc.
SHUNT				
Drops 3 to 5% from no load to full load	Any desired range depending on motor design and type of system.	Good. With constant field, varies directly as voltage applied to armature	High. Limited by commutation, heating and line capacity	Where constant or adjustable speed is required and starting conditions are not severe. Fan, blowers, centrifugal pumps, conveyers, wood working machines, metal working machines, elevators
COMPOUND				
Drops 7 to 20% from no load to full load depending on amount of compounding	Any desired range depending on motor design and type of control	Higher than for shunt, depending on amount of compounding	High. Limited by commutation, heating and line capacity	Where high starting torque combined with fairly constant speed is required. Plunger pumps, punch presses, shears, bending rolls, geared elevators, conveyors, hoists
SPLIT-PHASE				
Drops about 10% from no load to full load	None	75% for large to 175% for small sizes	150% for large to 200% for small sizes	Constant speed service where starting is easy, Small fans, centrifugal pumps, and light running machines, where polyphase current is not available

SUMMARY OF APPLICATIONS

DC and Single phase motors (cont'd)

Speed Regulation	Speed Control	Starting Torque	Pull-out Torque	Applications
CAPACITOR				
Drops 5% for large to 10% for small sizes	None	150 to 350% of full load depending upon design and size	150% for large to 200% for small sizes	Constant speed service for any starting duty, and quiet operation, where polyphase current cannot be used
COMMUTATOR-TYPE				
Drops 5% for large to 10% for small sizes	Repulsion-induction, none. Brush shifting types 4 to 1 at full load	250% for large to 350% for small sizes	150% for large to 250% for small sizes	Constant speed service for any starting duty, where speed control is required and polyphase current cannot be used

FOR MOTOR CHARACTERISTICS

2 and 3 phase motors

Speed Regulation	Speed Control	Starting Torque	Pull-out Torque	Applications
GENERAL-PURPOSE SQUIRREL-CAGE (Class B)				
Drops about 3% for large to 5% for small sizes	None, except multi-speed types designed for 2 to 4 fixed speeds	200% of full load for 2-pole to 105% for 16 pole designs	200% of full load	Constant-speed service where starting torque is not excessive fans, blowers, rotary compressers, centrifugal pumps
HIGH TORQUE SQUIRREL-CAGE (Class C)				
Drops about 3% for large to 6% for small sizes	None, except multi-speed types designed for 2 to 4 fixed speeds	250% of full load for high speed to 200% for low speed designs	200% of full load	Constant-speed service where fairly high starting torque is required at infrequent intervals with starting current of about 400% of full load. Reciprocating pumps and compressors, crushers, etc.
HIGH SLIP SQUIRREL-CAGE (Class D)				
Drops about 10 to 15% from no load to full load	None, except multi-speed types designed for 2 to 4 fixed speeds	225 to 300% of full load, depending on speed with rotor resistance	200% Will usually not stall until loaded to max. torque, which occurs at standstill	Constant-speed service and high starting torque, if starting is not too frequent, and for taking high peak loads with or without flywheels. Punch presses, shears and elevators, etc.
LOW TORQUE SQUIRREL-CAGE (Class F)				
Drops about 3% for large to 5% for small sizes	None, except multi-speed types designed for 2 to 4 fixed speeds	50% of full load for high speed to 90% for low speed designs	150 to 170% of full load	Constant speed service where starting duty is light. Fans blowers, centrifugal pumps and similar loads

FOR MOTOR CHARACTERISTICS

2 and 3 phase motors (cont'd)

Speed Regulation	Speed Control	Starting Torque	Pull-out Torque	Applications
WOUND-ROTOR				
With rotor rings short circuited, drops about 3% for large to 5% for small sizes	Speed can be reduced to 50% by rotor resistance to obtain stable operation. Speed varies inversely as load	Up to 300% depending on external resistance in rotor circuit and how distributed	200%. when rotor slip rings are short circuited	Where high starting torque with low starting current or where limited speed control is required. Fans, centrifugal and plunger pumps, compressers, conveyers, hoists, cranes
SYNCHRONOUS				
Constant	None, except special motors designed for 2 fixed speeds	40% for slow to 160% for medium speed 80% pf designs. Special designs develop higher torques	Unity-of motors 170%; 80% pf motors 225%. Special designs up to 300%	For constant-speed service, direct connection to slow speed machines and where power factor correction is required.

STARTING METHODS: SQUIRREL CAGE INDUCTION MOTORS

Starter Type	% Full-Voltage Value		
	Voltage at Motor	Line Current	Motor Output Torque
Full Voltage	100	100	100
Autotransformer			
80 pc tap	80	64*	64
65 pc tap	65	42*	42
50 pc tap	50	25*	25
Primary-reactor			
80 pc tap	80	80	64
65 pc tap	65	65	42
50 pc tap	50	50	25
Primary-resistor			
Typical rating	80	80	64
Part-winding			
Low-speed motors			
(1/2-1/2)	100	50	50
High-speed motors			
(1/2-1/2)	100	70	50
High-speed motors			
(2/3-1/3)	100	65	42
Wye Start-Delta Run			
(1/3-1/3)	100	33	33

*Autotransformer magnetizing current not included.
Magnetizing current usually less than 25 percent motor full-load current

SETTING BRANCH CIRCUIT PROTECTIVE DEVICES

Type of Motor	Percent of Full Load Current			
	Nontime Delay Fuse	Dual-Element (Time-Delay) Fuse	Instant. Trip Type Breaker	Time-Limit Breaker
All A.C. single-phase and polyphase squirrel-cage and synchronous motors with full-voltage, resistance or reactor starting:				
No code letter	300	175	700	250
Code letter F to V	300	175	700	250
Code letter B to E	250	175	700	200
Code letter A	150	150	700	150
All A.C. squirrel-cage and synchronous motors with auto-transformer starting:				
Code letter F to V	250	175	700	200
Code letter B to E	200	175	700	200
Code letter A	150	150	700	150
Wound Rotor	150	150	700	150
Direct Current				
Not more than 50 HP	150	150	250	150
More than 50 HP	150	150	175	150

GENERAL EFFECT OF VOLTAGE VARIATION ON DIRECT-CURRENT MOTOR CHARACTERISTICS

□ =INCREASE ● =DECREASE

Voltage Variation	Starting and Max Run Torque	Full-load Speed	EFFICIENCY			Full-load Current	Temperature Rise, Full Load	Maximum Overload Capacity	Magnetic Noise
			Full Load	3/4 Load	1/2 Load				
SHUNT-WOUND									
120% Voltage	□ 30%	110%	Slight □	No Change	Slight ●	● 17%	Main field □ Commutating field and armature ●	□ 30%	Slight □
110% Voltage	□ 15%	105%	Slight □	No Change	Slight ●	● 8.5%	Main field □ Commutating field and armature ●	□ 15%	Slight □
90% Voltage	● 16%	95%	Slight ●	No Change	Slight □	□ 11.5%	Main field ● Commutating field and armature □	● 16%	Slight ●
COMPOUND-WOUND									
120% Voltage	□ 30%	112%	Slight □	No Change	Slight ●	● 17%	Main field □ Commutating field and armature ●	□ 30%	Slight □
110% Voltage	□ 15%	106%	Slight □	No Change	Slight ●	● 8.5%	Main field □ Commutating field and armature ●	□ 15%	Slight □
90% Voltage	● 16%	94%	Slight ●	No Change	Slight □	□ 11.5%	Main field ● Commutating field and armature □	● 16%	Slight ●

NOTES:--Starting current is controlled by starting resistor.
This table shows general effects, which will vary somewhat for specific ratings.

GENERAL EFFECT OF VOLTAGE AND FREQUENCY VARIATION ON INDUCTION-MOTOR CHARACTERISTICS

NOTE: This table shows general effects, which will vary somewhat for specific ratings.   □ =INCREASE   ● =DECREASE

		Starting and Max Running Torque	Synchronous Speed	% Slip	Full-load Speed	Efficiency Full Load	Efficiency 3/4 Load	Efficiency 1/2 Load	Power-Factor Full Load	Power-Factor 3/4 Load	Power-Factor 1/2 Load	Full-load Current	Starting Current	Temperature Rise, Full Load	Maximum Overload Capacity	Magnetic Noise, No Load in Particular
Voltage Variation	120% Voltage	□ 44%	No Change	● 30%	□ 1.5%	Small □	□ 1/2 to 2 points	● 7 to 20 points	● 5 to 15 points	● 10 to 30 points	● 15 to 40 points	● 11%	□ 25%	● 5 to 6C	□ 44%	Noticeable □
	110% Voltage	□ 21%	No Change	● 17%	□ 1%	□ 1/2 to 1 point	Practically no change	● 1 to 2 points	● 3 points	● 4 points	● 5 to 6 points	● 7%	□ 10 to 12%	● 3 to 4C	□ 21%	Slight □
	Function of Voltage	$(\text{Voltage})^2$ □	Constant	$\dfrac{1}{(\text{Voltage})^2}$ □	(Syn speed slip)	—	—	—	—	—	—	—	Voltage	—	$(\text{Voltage})^2$ □	—
	90% Voltage	● 19%	No Change	□ 23%	□ 1½%	● 2 points	Practically no change	□ 1 to 2 points	□ 1 point	□ 2 to 3 points	□ 4 to 5 points	□ 11%	● 10 to 12%	□ 6 to 7C	● 19%	Slight ●
Frequency Variation	105% Frequency	● 10%	□ 5%	Practically no change	□ 5%	Slight □	Slight □	Slight □	Slight □	Slight □	Slight □	Slight ●	● 5 to 6%	Slight □	Slight □	Slight ●
	Function of Frequency	$\dfrac{1}{(\text{Frequency})^2}$ ●	Frequency	Practically no change	(Syn speed slip)	—	—	—	—	—	—	—	$\dfrac{1}{(\text{Frequency})}$ ●	—	—	—
	95% Frequency	□ 11%	● 5%	Practically no change	● 5%	Slight ●	Slight ●	Slight ●	Slight ●	Slight ●	Slight ●	Slight □	□ 5 to 6%	Slight □	—	Slight □

6-23

CONTACTOR AND MOTOR STARTER TROUBLESHOOTING GUIDE

Problem	Possible Cause	Corrective Action
Humming noise	Magnet pole faces misaligned	Realign, Replace magnet assembly if realignment is not possible.
	Too low voltage at coil	Measure voltage at coil. Check voltage rating of coil. Correct any voltage that is 10% less than coil rating
	Pole face obstructed by foreign object, dirt, or rust	Remove any foreign object and clean as necessary. Never file pole faces
Loud buzz noise	Shading coil broken	Replace coil assembly.
Controller fails to drop out	Voltage to coil not being removed	Measure voltage at coil. Trace voltage from coil to supply looking for short-ed switch or contact if voltage is present.
	Worn or rusted parts causing binding	Clean rusted parts, Replace worn parts.
	Contact poles sticking	Checking for burning or sticky substance on contacts. Replace burned con-tacts. Clean dirty contacts
	Mechanical interlock binding	Check to ensure interlocking mechanism is free to move when power is OFF. Replace faulty interlock.
Controller fails to pull in	No coil voltage	Measure voltage at coil terminals. Trace voltage loss from coil to supply voltage if voltage is not present.
	Too low voltage	Measure voltage at coil terminals. Correct voltage level if voltage is less than 10% of rated coil voltage. Check for a voltage drop as large loads are energized
	Coil open	Measure voltage at coil. Remove coil if voltage is present and correct but coil does not pull in. Measure coil resistance for open circuit. Replace if open.

6-24

CONTACTOR AND MOTOR STARTER TROUBLESHOOTING GUIDE

Problem	Possible Cause	Corrective Action
Contacts badly burned or welded	Coil shorted	Shorted coil may show signs of burning. The fuse or breakers should trip if coil is shorted. Disconnect one side of coil and reset if tripped. Remove coil and check resistance for short if protection device does not trip. Replace shorted coil. Replace any coil that is burned.
	Mechanical obstruction	Remove any obstructions.
	Too high inrush current	Measure inrush current. Check load for problem if higher-than-rated load current. Change to larger controller if load current is correct but excessive for controller.
	Too fast load cycling	Change to larger controller if load cycled ON and OFF repeatedly.
	Too large overcurrent protection device	Size overcurrent protection to load and controller.
	Short circuit	Check fuses or breakers. Clear any short circuit.
	Insufficient contact pressure	Check to ensure contacts are making good connection.
Nuisance tripping	Incorrect overload size	Check size of overload against rated load current. Size up if permissible per NEC®.
	Lack of temperature compensation	Correct setting of overload if controller and load are at different ambient temperatures
	Loose connections	Check for loose terminal connection

DIRECT CURRENT MOTOR TROUBLESHOOTING GUIDE

Problem	Possible Cause	Corrective Action
Motor will not start	Blown fuse or open CB	Test the OCPD. If voltage is present at the input, but not the output of the OCPD, the fuse is blown or the CB is open. Check the rating of the OCPD. It should be at least 125% of the motor's FLC.
	Motor overload on starter tripped.	Allow overloads to cool. If reset overloads do not start motor, test the starter.
	No brush contact	Check brushes. Replace, if worn.
	Open control circuit between incoming power and motor	Check for cleanliness, tightness, and breaks. Use a voltmeter to test the circuit starting with the incoming power and moving to the motor terminals. Voltage generally stops at the problem area.
Fuse, CB, or overloads retrip after service	Excessive load	If the motor is loaded to excess or is jammed, the circuit OCPD will open. Disconnect the load from the motor. If the motor now runs properly, check the load. If the motor does not run and the fuse or CB opens, the problem is with the motor or control circuit. Remove the motor from the control circuit and connect it directly to the power source. If the motor runs properly, the problem is in the control circuit. Check the control circuit. If the motor opens the fuse or CB again, the problem is in the motor. Replace or service the motor.
	Motor shaft does not turn	Disconnect the motor from the load. If the motor shaft still does not turn, the bearings are frozen. Replace or service the motor.
Brushes chip or break	Brush material is too weak or the wrong type for motor's duty rating	Replace with better grade or type of brush. Consult manufacturer if problem continues.
	Brush face is overheating and losing brush bonding material.	Check for an overload on the motor. Reduce the load as required. Adjust brush holder arms.

DIRECT CURRENT MOTOR TROUBLESHOOTING GUIDE

Problem	Possible Cause	Corrective Action
Brushes chip or break.	Brush holder is too far from commutator.	Too much space between the brush holder and the surface of the commutator allows the brush end to chip or break. Set correct space between brush holder and commutator.
	Brush tension is incorrect	Adjust brush tension so the brush rides freely on the commutator.
Brushes spark.	Worn brushes	Replace worn brushes. Service the motor if rapid brush wear, excessive sparking, chipping, breaking, or chattering is present.
	Commutator is concentric.	Grind commutator and undercut mica. Replace commutator if necessary.
	Excessive vibration	Balance armature. Check brushes. They should be riding freely.
Rapid brush wear	Wrong brush material, type, or grade	Replace with brushes recommended by manufacturer.
	Incorrect brush tension	Adjust brush tension so the brush rides freely on the commutator
Motor overheats.	Improper ventilation	Clean all ventilation openings. Vacuum or blow dirt out of motor with low-pressure, dry, compressed air.
	Motor is overloaded	Check the load for binding. Check shaft straightness. Measure motor current under operating conditions. If the current is above the listed current rating, remove the motor. Remeasure the current under no-load conditions. If the current is excessive under load but not when unloaded, check the load. If the motor draws excessive current when disconnected, replace or service the motor.

SHADED POLE MOTOR TROUBLE SHOOTING GUIDE

Problem	Possible Cause	Corrective Action
Motor will not start.	Blown fuse or open CB	Test OCPD. If voltage is present at the input, but not the output of the OCPD, the fuse is blown or the CB is open. Check the rating of the OCPD. It should be at least 125% of the motor's FLC.
	Motor overload on starter tripped.	Allow overloads to cool. Reset overloads. If reset overloads do not start the motor, test the starter.
	Low or no voltage applied to motor	Check the voltage at the motor terminals. The voltage must be present and within 10% of the motor nameplate voltage. If voltage is present at the motor but the motor is not operating, remove the motor from the load the motor is driving. Reapply power to the motor. If the motor runs, the problem is with the load. If the motor does not run, the problem is with the motor. Replace or service the motor.
	Open control circuit between incoming power and motor	Check for cleanliness, tightness, and breaks. Use a voltmeter to test the circuit starting with the incoming power and moving to the motor terminals. Voltage generally stops at the problem area.
Fuse, CB, or overloads retrip after service	Excessive load	If the motor is loaded to excess or jammed, the circuit OCPD will open. Disconnect the load from the motor. If the motor now runs properly, check the load. If the motor does not run and the fuse or CB opens, the problem is with the motor or control circuit. Remove the motor from the control circuit and connect it directly to the power source. If the motor runs properly, the problem is in the control circuit. Check the control circuit, if the motor opens the fuse or CB again, the problem is in the motor. Replace or service the motor.
Excessive noise	Unbalanced motor or load	An unbalanced motor or load causes vibration, which causes noise. Realign the motor and load. Check for excessive end play or loose parts. If the shaft is bent, replace the rotor or motor.
	Dry or worn bearings	Dry or worn bearings cause noise. Bearings may be dry due to dirty oil, oil not reaching the shaft, or motor overheating. Oil bearings as recommended. If noise remains, replace the bearings or motor.
	Excessive grease	Ball bearings that have excessive grease may cause bearings to overheat. Overheated bearings cause noise. Remove excess grease.

SPLIT-PHASE MOTOR TROUBLESHOOTING GUIDE

Problem	Possible Cause	Corrective Action
Motor will not start	Thermal cutout switch is open	Reset the thermal switch. Caution: Resetting the thermal switch may automatically start the motor.
	Blown fuse or open CB	Test the OCPD, if voltage is present at the input, but not the output of the OCPD, the fuse is blown or the CB is open. Check the rating of the OCPD. It should be at least 125% of the motor's FLC.
	Motor overload on starter tripped.	Allow overloads to cool. Reset overloads. If reset overloads do not start the motor, test the starter.
	Low or no voltage applied to motor	Check the voltage at the motor terminals. The voltage must be present and within 10% of the motor nameplate voltage. If voltage is present at the motor but the motor is not operating, remove the motor from the load the motor is driving. Reapply power to the motor. If the motor runs, the problem is with the load. If the motor does not run, the problem is with the motor. Replace or service the motor.
	Open control circuit between incoming power and motor	Check for cleanliness, tightness, and breaks. Use a voltmeter to test the circuit starting with the incoming power and moving to the motor terminals. Voltage generally stops at the problem area
	Starting winding not receiving power	Check the centrifugal switch to make sure it connects the starting winding when the motor is OFF.
Fuse, CB, or overloads retrip after service	Blown fuse or open CB	Test the OCPD. If voltage is present at the input, but not the output of the OCPD, the fuse is blown or the CB is open. Check the rating of the OCPD. It should be at least 125% of the motor's FLC.
	Motor overload on starter tripped.	Allow overloads to cool. Reset overloads. If reset overloads do not start the motor, test the starter.

SPLIT-PHASE MOTOR TROUBLESHOOTING GUIDE

Problem	Possible Cause	Corrective Action
Fuse, CB, or overloads retrip after service	Low or no voltage applied to motor	Check the voltage at the motor terminals. The voltage must be present and within 10% of the motor nameplate voltage. If voltage is present at the motor but the motor is not operating, remove the motor from the load the motor is driving. Reapply power to the motor. If the motor runs, the problem is with the load. If the motor does not run, the problem is with the motor. Replace or service the motor.
	Open control circuit between incoming power and motor	Check for cleanliness, tightness, and breaks. Use a voltmeter to test the circuit starting with the incoming power and moving to the motor terminals. Voltage generally stops at the problem area.
	Motor shaft does not turn.	Disconnect the motor from the load. If the motor shaft still does not turn, the bearings are frozen. Replace or service the motor.
Motor produces electric shock	Broken or disconnected ground strap	Connect or replace ground strap. Test for proper ground.
	Hot power lead at motor connecting terminals is touching motor frame.	Disconnect the motor. Open the motor terminal box and check for poor connections, damaged insulation, or leads touching the frame. Service and test motor for ground.
	Motor winding shorted to frame.	Remove, service, and test motor.
Motor overheats.	Starting windings are not being removed from circuit as motor accelerates	When the motor is turned OFF, a distinct click should be heard as the centrifugal switch closes.
	Improper ventilation	Clean all ventilation openings. Vacuum or blow dirt out of motor with low-pressure, dry, compressed air.

SPLIT PHASE MOTOR TROUBLESHOOTING GUIDE

Problem	Possible Cause	Corrective Action
Motor overheats.	Motor is overloaded	Check the load for binding. Check shaft straightness. Measure motor current under operating conditions. If current is above the listed current rating, remove the motor. Remeasure the current under no-load conditions. If the current is excessive under load but not when unloaded, check the load. If the motor draws excessive current when disconnected, replace or service the motor.
	Dry or worn bearings	Dry or worn bearings cause noise. The bearings may be dry due to dirty oil, oil not reaching the shaft, or motor overheating. Oil the bearings as recommended. If noise remains, replace the bearings or the motor.
	Dirty bearings	Clean or replace bearings.
Excessive noise	Excessive end-play	Check end play by trying to move the motor shaft in and out. Add end-play washers as required.
	Unbalanced motor or load	An unbalanced motor or load causes vibration, which causes noise. Realign the motor and load. Check for excessive end play or loose parts. If the shaft is bent, replace the rotor or motor.
	Dry or worn bearings	Dry or worn bearings cause noise. The bearings may be dry due to dirty oil, oil not reaching the shaft, or motor overheating. Oil the bearings as recommended. If noise remains, replace the bearings or the motor.
	Excessive grease	Ball bearings that have excessive grease may cause the bearings to overheat. Overheated bearings cause noise. Remove any excess grease.

THREE-PHASE MOTOR TROUBLESHOOTING GUIDE

Problem	Possible Cause	Corrective Action
Motor will not start	Wrong motor connections	Most 3φ motors are dual-voltage. Check for proper motor connections.
	Blown fuse or open CB	Test the OCPD. If voltage is present at the input, but not the output of the OCPD, the fuse is blown or the CB is open. Check the rating of the OCPD. It should be at least 125% of the motor's FLC.
	Motor overload on starter tripped	Allow overloads to cool. Reset overloads. If reset overloads do not start the motor, test the starter.
	Low or no voltage applied to motor	Check the voltage at the motor terminals. The voltage must be present and within 10% of the motor nameplate voltage. If voltage is present at the motor but the motor is not operating, remove the motor from the load. Reapply power to the motor. If the motor runs, the problem is with the load. If the motor does not run, the problem is with the motor. Replace or service the motor.
	Open control circuit between incoming power and motor	Check for cleanliness, tightness, and breaks. Use a voltmeter to test the circuit starting with the incoming power and moving to the motor terminals. Voltage generally stops at the problem area.
Fuse, CB, or overloads retrip after service.	Power not applied to all three lines	Measure voltage at each power line. Correct any power supply problems.
	Blown fuse or open CB	Test the OCPD. If voltage is present at the input, but not the output of the OCPD, the fuse is blown or the CB is open. Check the rating of the OCPD. It should be at least 125% of the motor's FLC.
	Motor overload on starter tripped.	Allow overloads to cool. Reset overloads. If reset overloads do not start the motor, test the starter.

THREE-PHASE MOTOR TROUBLESHOOTING GUIDE

Problem	Possible Cause	Corrective Action
Fuse, CB, or overloads retrip after service.	Low or no voltage applied to motor	Check the voltage at the motor terminals. The voltage must be present and within 10% of the motor nameplate voltage. If voltage is present at the motor but the motor is not operating, remove the motor from the load the motor is driving. Reapply power to the motor. If the motor runs, the problem is with the load. If the motor does not run, the problem is with the motor. Replace or service the motor.
	Open control circuit between incoming power and motor	Check for cleanliness, tightness, and breaks. Use a voltmeter to test the circuit starting with the incoming power and moving to the motor terminals. Voltage generally stops at the problem area.
	Motor shaft does not turn.	Disconnect the motor from the load. If the motor shaft still does not turn, the bearings are frozen. Replace or service the motor.
Motor overheats	Motor is single phasing.	Check each of the 3φ power lines for correct voltage.
	Improper ventilation	Clean all ventilation openings. Vacuum or blow dirt out of motor with low-pressure, dry, compressed air.
	Motor is overloaded	Check the load for binding. Check shaft straightness. Measure motor current under operating conditions. If the current is above the listed current rating, remove the motor. Remeasure the current under no-load conditions. If the current is excessive under load but not when unloaded, check the load. If the motor draws excessive current when disconnected, replace or service the motor.

FULL-LOAD CURRENTS
DC MOTORS

Motor rating (HP)	Current (A)	
	120V	240V
1/4	3.1	1.6
1/3	4.1	2.0
1/2	5.4	2.7
3/4	7.6	3.8
1	9.5	4.7
1 1/2	13.2	6.6
2	17	8.5
3	25	12.2
5	40	20
7 1/2	48	29
10	76	38

FULL-LOAD CURRENTS
AC MOTORS

Motor rating (HP)	Current (A)	
	115V	230V
1/6	4.4	2.2
1/4	5.8	2.9
1/3	7.2	3.6
1/2	9.8	4.9
3/4	13.8	6.9
1	16	8
1 1/2	20	10
2	24	12
3	34	17
5	56	28
7 1/2	80	40

FULL-LOAD CURRENTS — 3φ, AC INDUCTION MOTORS

Motor rating (HP)	Current (A)			
	208V	230V	460V	575V
1/4	1.11	.96	.48	.38
1/3	1.34	1.18	.59	.47
1/2	2.2	2.0	1.0	.8
3/4	3.1	2.8	1.4	1.1
1	4.0	3.6	1.8	1.4
1 1/2	5.7	5.2	2.6	2.1
2	7.5	6.8	3.4	2.7
3	10.6	9.6	4.8	3.9
5	16.7	15.2	7.6	6.1
7 1/2	24.0	22.0	11.0	9.0
10	31.0	28.0	14.0	11.0
15	46.0	42.0	21.0	17.0
20	59	54	27	22
25	75	68	34	27
30	88	80	40	32
40	114	104	52	41
50	143	130	65	52
60	169	154	77	62
75	211	192	96	77
100	273	248	124	99
125	343	312	.156	125
150	396	360	180	144
200	—	480	240	192
250	—	602	301	242
300	—	—	362	288
350	—	—	413	337
400	—	—	477	382
500	—	—	590	472

TYPICAL MOTOR EFFICIENCIES

HP	Standard Motor (%)	Energy-Efficient Motor (%)
1	76.5	84.0
1.5	78.5	85.5
2	79.9	86.5
3	80.8	88.5
5	83.1	88.6
7.5	83.8	90.2
10	85.0	90.3
15	86.5	91.7
20	87.5	92.4
25	88.0	93.0
30	88.1	93.1
40	89.3	93.6
50	90.4	93.7
75	90.8	95.0
100	91.6	95.4
125	91.8	95.8
150	92.3	96.0
200	93.3	96.1
250	93.6	96.2
300	93.8	96.5

1φ MOTORS AND CIRCUITS

1		2		3	4	5				6	
Size of motor		Motor overload protection Low-peak or Fusetron*		Switch 115% minimum or HP rated or fuse holder size	Minimum size of starter	Controller termination temperature rating				Minimum size of copper wire and trade conduit	
		Motor less than 40°C or greater than 1.15 SF (Max fuse 125%)	All other motors (Max fuse 115%)			60°C		75°C		Wire size (AWG or kcmil)	Conduit (inches)
HP	Amp					TW	THW	TW	THW		
115 V (120 V system)											
1/6	4.4	5	5	30	00	•	•	•	•	14	1/2
1/4	5.8	7	6 1/4	30	00	•	•	•	•	14	1/2
1/3	7.2	9	8	30	00	•	•	•	•	14	1/2
1/2	9.8	12	10	30	00	•	•	•	•	14	1/2
3/4	13.8	15	15	30	00	•	•	•	•	14	1/2
1	16	20	17 1/2	30	00	•	•	•	•	14	1/2
1 1/2	20	25	20	30	01	•	•	•	•	12	1/2
2	24	30	25	30	01	•	•	•	•	10	1/2

1φ MOTORS AND CIRCUITS

1		2		3	4	5				6	
Size of motor		Motor overload protection Low-peak or Fusetron*		Switch 115% minimum or HP rated or fuse holder size	Minimum size of starter	Controller termination temperature rating				Minimum size of copper wire and trade conduit	
		Motor less than 40°C or greater than 1.15 SF (Max fuse 125%)	All other motors (Max fuse 115%)			60°C		75°C		Wire size (AWG or kcmil)	Conduit (inches)
HP	Amp					TW	THW	TW	THW		
230 V (240 V system)											
1/6	2.2	2 1/2	2 1/2	30	00	•	•	•	•	14	1/2
1/4	2.9	3 1/2	3 2/10	30	00	•	•	•	•	14	1/2
1/3	3.6	4 1/2	4	30	00	•	•	•	•	14	1/2
1/2	4.9	5 6/10	5 6/10	30	00	•	•	•	•	14	1/2
3/4	6.9	8	7 1/2	30	00	•	•	•	•	14	1/2
1	8	10	9	30	00	•	•	•	•	14	1/2
1 1/2	10	12	10	30	00	•	•	•	•	14	1/2
2	12	15	12	30	0	•	•	•	•	14	1/2
3	17	20	17 1/2	30	1	•	•	•	•	12	1/2

1	2		3	4	5				6	
Size of motor	Motor overload protection Low-peak or Fusetron*		Switch 115% minimum or HP rated or fuse holder size	Minimum size of starter	Controller termination temperature rating				Minimum size of copper wire and trade conduit	
	Motor less than 40°C or greater than 1.15 SF (Max fuse 125%)	All other motors (Max fuse 115%)			60°C		75°C		Wire size (AWG or kcmil)	Conduit (inches)
HP	Amp				TW	THW	TW	THW		
230 V (240 V system)										
5	28	35	30*	60	2		•			8	3/4
								•	8	1/2
7 1/2	40	50	45	60	2				•	10	1/2
						•			6	3/4
								•	8	3/4
10	50	60	50	60	3		•			4	1
								•	6	3/4

* Fuse reducers required.

3φ 230 V MOTORS AND CIRCUITS—240 V SYSTEM

	1	2	3	4	5				6	
	Size of motor		Motor overload protection Low-peak or Fusetron*	Switch 115% minimum or HP rated or fuse holder size	Minimum size of starter	Controller termination temperature rating				Minimum size of copper wire and trade conduit	
						60°C		75°C			
HP	Amp	Motor less than 40°C or greater than 1.15 SF (Max fuse 125%)	All other motors (Max fuse 115%)			TW	THW	TW	THW	Wire size (AWG or kcmil)	Conduit (inches)
1/2	2	2 1/2	2 1/4	30	00	•	•	•	•	14	1/2
3/4	2.8	3 1/2	3 2/10	30	00	•	•	•	•	14	1/2
1	3.6	4 1/2	4	30	00	•	•	•	•	14	1/2
1 1/2	5.2	6 1/4	5 6/10	30	00	•	•	•	•	14	1/2
2	6.8	8	7 1/2	30	0	•	•	•	•	14	1/2
3	9.6	12	10	30	0	•	•	•	•	14	1/2
5	15.2	17 1/2	17 1/2	30	1	•	•	•	•	14	1/2
7 1/2	22	25	25	30	1	•	•	•	•	10	1/2
10	28	35	30*	60	2	•		•		8	3/4
								•	•	10	1/2
15	42	50	45	60	2	•				6	1
								•	•	6	3/4

*Fuse reducers required.

3φ 230 V MOTORS AND CIRCUITS—240 V SYSTEM

1		2		3	4	5				6	
Size of motor		Motor overload protection Low-peak or Fusetron*		Switch 115% minimum or HP rated or fuse holder size	Minimum size of starter	Controller termination temperature rating				Minimum size of copper wire and trade conduit	
						60°C		75°C			
HP	Amp	Motor less than 40°C or greater than 1.15 SF (Max fuse 125%)	All other motors (Max fuse 115%)			TW	THW	TW	THW	Wire size (AWG or kcmil)	Conduit (inches)
20	54	60*	60*	100	3	•	•			4	1
								•	•	3	1 1/4
25	68	80	75	100	3	•				3	1
								•		4	1
30	80	100	90	100	3	•				1	1 1/4
								•		3	1 1/4
40	104	125	110	200	4	•				2/0	1 1/2
								•		1	1 1/4
50	130	150	150	200	4	•				3/0	2
								•		2/0	1 1/2

Fuse reducers required.

6-41

3φ 230 V MOTORS AND CIRCUITS—240 V SYSTEM

1		2		3	4	5				6	
Size of motor		Motor overload protection Low-peak or Fusetron*		Switch 115% minimum or HP rated or fuse holder size	Minimum size of starter	Controller termination temperature rating				Minimum size of copper wire and trade conduit	
		Motor less than 40°C or greater than 1.15 SF (Max fuse 125%)	All other motors (Max fuse 115%)			60°C			75°C		
HP	Amp					TW	THW	TW	THW	Wire size (AWG or kcmil)	Conduit (inches)
75	192	225	200*	400	5	•	•		•	300	2 1/2
									•	250	2 1/2
100	248	300	250	400	5	•	•		•	500	3
									•	350	2 1/2
150	360	450	400*	600	6	•	•			300-2φ	2-2 1/2
									•	4/0-2φ	2-2

* Fuse reducers required.

3φ 460 V MOTORS AND CIRCUITS—480 V SYSTEM

1	2			3	4	5				6	
Size of motor	Motor overload protection Low-peak or Fusetron*			Switch 115% minimum or HP rated or fuse holder size	Minimum size of starter	Controller termination temperature rating				Minimum size of copper wire and trade conduit	
		Motor less than 40°C or greater than 1.15 SF (Max fuse 125%)	All other motors (Max fuse 115%)			60°C		75°C			
HP	Amp					TW	THW	TW	THW	Wire size (AWG or kcmil)	Conduit (inches)
1/2	1	1 1/4	1 1/8	30	00	•				14	1/2
3/4	1.4	1 6/10	1 6/10	30	00	•				14	1/2
1	1.8	2 1/4	2	30	00	•				14	1/2
1 1/2	2.6	3 2/10	2 6/10	30	00	•				14	1/2
2	3.4	4	3 1/2	30	00	•				14	1/2
3	4.8	5 6/10	5	30	0	•				14	1/2
5	7.6	9	8	30	0	•				14	1/2
7 1/2	11	12	12	30	1	•				14	1/2
10	14	17 1/2	15	30	1	•				14	1/2
15	21	25	20	30	2		•			10	1/2
20	27	30*	30*	60	2			•		8	3/4
									•	10	1/2

* Fuse reducers required.

3φ 460 V MOTORS AND CIRCUITS—480 V SYSTEM

1		2		3	4	5				6	
Size of motor		Motor overload protection Low-peak or Fusetron*		Switch 115% minimum or HP rated or fuse holder size	Minimum size of starter	Controller termination temperature rating				Minimum size of copper wire and trade conduit	
		Motor less than 40°C or greater than 1.15 SF (Max fuse 125%)	All other motors (Max fuse 115%)			60°C		75°C			
HP	Amp					TW	THW	TW	THW	Wire size (AWG or kcmil)	Conduit (inches)
25	34	40	35	60	2	•	•			6	1
								•	•	8	3/4
30	40	50	45	60	3	•	•			6	1
								•	•	8	3/4
40	52	60*	60*	100	3	•	•			4	1
								•	•	6	1
50	65	80	70	100	3	•	•			3	1 1/4
								•	•	4	1
60	77	90	80	100	4	•	•			1	1 1/4
								•	•	3	1 1/4
75	96	110	110	200	4	•	•			1/0	1 1/2
								•	•	1	1 1/4

* Fuse reducers required.

3Φ 460 V MOTORS AND CIRCUITS—480 V SYSTEM

1		2		3	4	5				6	
Size of motor		Motor overload protection Low-peak or Fusetron*		Switch 115% minimum or HP rated or fuse holder size	Minimum size of starter	Controller termination temperature rating				Minimum size of copper wire and trade conduit	
		Motor less than 40°C or greater than 1.15 SF (Max fuse 125%)	All other motors (Max fuse 115%)			60°C		75°C		Wire size (AWG or kcmil)	Conduit (inches)
HP	Amp					TW	THW	TW	THW		
100	124	150	125	200	4	•	•			3/0	2
									•	2/0	1-1/2
125	156	175	175	200	5	•	•			4/0	2
									•	3/0	2
150	180	225	200*	400	5	•	•			300	2-1/2
									•	4/0	2
200	240	300	250	400	5	•	•			500	3
									•	350	2-1/2
250	302	350	325	400	6	•	•			4/0-2φ	2-2
									•	3/0-2φ	2-2
300	361	450	400*	600	6	•	•			300-2φ	2—1-1/2
									•	4/0-2φ	2-2

* Fuse reducers required.

STANDARD SIZES CF FUSES AND CB'S

NEC* 240-6 lists standard ampere ratings of fuses and fixed-trip CB's as follows:
15, 20, 25, 30, 35, 40, 45,
50, 60, 70, 80, 90, 100, 110,
125, 150, 175, 200, 225,
250, 300, 350, 400, 450,
500, 600, 700, 800,
1000, 1200, 1600,
2000, 2500, 3000, 4000, 5000, 6000

OVERCURRENT PROTECTION DEVICES

Motor Type	Code Letter	FLC (%)				
		Motor Size	TDF	NTDF	ITB	ITCB
AC*	—	—	175	300	150	700
AC*	A	—	150	150	150	700
AC*	B—E	—	175	250	200	700
AC*	F—V	—	175	300	250	700
DC	—	1/8 to 50 HP	150	150	150	250
DC	—	Over 50 HP	150	150	150	175

* full-voltage and resistor starting

DC MOTORS AND CIRCUITS

1	2	3		4	5					6	
Size of motor		Motor overload protection — Dual-element fuse			Controller termination temperature rating					Minimum size of copper wire and trade conduit	
					60°C			75°C			
		Motor less than 40°C or greater than 1.15 SF (Max fuse 125%)	All other motors (Max fuse 115%)		TW	THW	TW	THW	THW		
HP	Amp			Switch 115% minimum or HP rated or fuse holder size						Wire size (AWG or kcmil)	Conduit (inches)
					Minimum size of starter						
90 V											
1/4	4.0	5	4 1/2	30	0	•	•	•	•	•	14	1/2
1/3	5.2	6 1/4	5 6/10	30	0	•	•	•	•	•	14	1/2
1/2	6.8	8	7.5	30	0	•	•	•	•	•	14	1/2
3/4	9.6	12	10	30	0	•	•	•	•	•	14	1/2
1	12.2	15	12	30	0	•	•	•	•	•	14	1/2
120 V											
1/4	3.1	3 1/2	3 1/2	30	0	•	•	•	•	•	14	1/2
1/3	4.1	5	4 1/2	30	0	•	•	•	•	•	14	1/2
1/2	5.4	6 1/4	6	30	0	•	•	•	•	•	14	1/2
3/4	7.6	9	8	30	0	•	•	•	•	•	14	1/2
1	9.5	10	10	30	0	•	•	•	•	•	14	1/2

* Fuse reducers required.

DC MOTORS AND CIRCUITS

1		2		3	4	5					6	
Size of motor		Motor overload protection — Dual-element fuse		Switch 115% minimum or HP rated or fuse holder size	Minimum size of starter	Controller termination temperature rating					Minimum size of copper wire and trade conduit	
						60°C			75°C			
HP	Amp	Motor less than 40°C or greater than 1.15 SF (Max fuse 125%)	All other motors (Max fuse 115%)			TW	THW	TW	TW	THW	Wire size (AWG or kcmil)	Conduit (inches)
120 V Continued												
1 1/2	13.2	15	15	30	1	•	•	•		•	14	1/2
2	17	20	17 1/2	30	1	•	•	•		•	12	1/2
5	40	50	45	60	2	•	•			•	6	3/4
											8	3/4
10	76	90	80	100	3	•	•			•	2	1
											3	1
180 V												
1/4	2	2 1/2	2 1/4	30	0	•	•	•	•	•	14	1/2
1/3	2.6	3 2/10	2 8/10	30	0	•	•	•	•	•	14	1/2
1/2	3.4	4	3 1/2	30	0	•	•	•	•	•	14	1/2
3/4	4.8	6	5	30	0	•	•	•	•	•	14	1/2

* Fuse reducers required.

DC MOTORS AND CIRCUITS

1		2		3	4	5				6	
Size of motor		Motor overload protection Dual-element fuse		Switch 115% minimum or HP rated or fuse holder size	Minimum size of starter	Controller termination temperature rating				Minimum size of copper wire and trade conduit	
						60°C			75°C		
HP	Amp	Motor less than 40°C or greater than 1.15 SF (Max fuse 125%)	All other motors (Max fuse 115%)			TW	THW	TW	THW	Wire size (AWG or kcmil)	Conduit (inches)
180 V Continued											
1	6.1	7 1/2	7	30	0	•	•	•		14	1/2
1 1/2	8.3	10	9	30	1	•	•	•		14	1/2
2	10.8	12	12	30	1	•	•	•		14	1/2
3	16	20	17 1/2	30	1	•	•	•		12	1/2
5	27	30*	30*	60	1	•				8	1/2
									•	**10**	**3/4**

* Fuse reducers required.

6-49

HORSEPOWER TO TORQUE CONVERSION

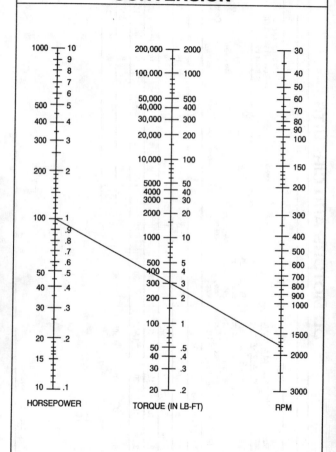

HEATER TRIP CHARACTERISTICS

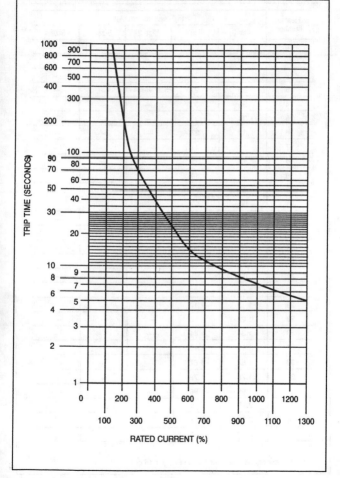

HEATER SELECTIONS

Heater number	Full-load current (A)*				
	Size 0	Size 1	Size 2	Size 3	Size 4
10	.20	.20	—	—	—
11	.22	.22	—	—	—
12	.24	.24	—	—	—
13	.27	.27	—	—	—
14	.30	.30	—	—	—
15	.34	.34	—	—	—
16	.37	.37	—	—	—
17	.41	.41	—	—	—
18	.45	.45	—	—	—
19	.49	.49	—		
20	.54	.54	—	—	—
21	.59	.59	—	—	—
22	.65	.65	—	—	—
23	.71	.71	—	—	—
24	.78	.78	—	—	—
25	.85	.85	—	—	—
26	.93	.93	—	—	—
27	1.02	1.02	—	—	—
28	1.12	1.12	—	—	—
29	1.22	1.22	—	—	—
30	1.34	1.34	—	—	—
31	1.48	1.48	—	—	—
32	1.62	1.62	—	—	—
33	1.78	1.78	—	—	—
34	1.96	1.96	—	—	—
35	2.15	2.15	—	—	—
36	2.37	2.37	—	—	—
37	2.60	2.60	—	—	—

*Full-load current (A) does not include FLC x 1.15 or 1.25.

HEATER SELECTIONS Continued

Heater number	Full-load current (A)*				
	Size 0	Size 1	Size 2	Size 3	Size 4
38	2.86	2.86	—	—	—
39	3.14	3.14	—	—	—
40	3.45	3.45	—	—	—
41	3.79	3.79	—	—	—
42	4.17	4.17	—	—	—
43	4.58	4.58	—	—	—
44	5.03	5.03	—	—	—
45	5.53	5.53	—	—	—
46	6.08	6.08	—	—	—
47	6.68	6.68	—	—	—
48	7.21	7.21	—	—	—
49	7.81	7.81	7.89	—	—
50	8.46	8.46	8.57	—	—
51	9.35	9.35	9.32	—	—
52	10.00	10.00	10.1	—	—
53	10.7	10.7	11.0	12.2	—
54	11.7	11.7	12.0	13.3	—
55	12.6	12.6	12.9	14.3	—
56	13.9	13.9	14.1	15.6	—
57	15.1	15.1	15.5	17.2	—
58	16.5	16.5	16.9	18.7	—
59	18.0	18.0	18.5	20.5	—
60	—	19.2	20.3	22.5	23.8
61	—	20.4	21.8	24.3	25.7
62	—	21.7	23.5	26.2	27.8
63	—	23.1	25.3	28.3	30.0
64	—	24.6	27.2	30.5	32.5
65	—	26.2	29.3	33.0	35.0

*Full-load current (A) does not include FLC x 1.15 or 1.25.

HEATER SELECTIONS Continued

Heater number	Full-load current (A)*				
	Size 0	Size 1	Size 2	Size 3	Size 4
66	—	27.8	31.5	36.0	38.0
67	—	—	33.5	39.0	41.0
68	—	—	36.0	42.0	44.5
69	—	—	38.5	45.5	48.5
70	—	—	41.0	49.5	52
71	—	—	43.0	53	57
72	—	—	46.0	58	61
73	—	—	—	63	67
74	—	—	—	68	72
75	—	—	—	73	77
76	—	—	—	78	84
77	—	—	—	83	91
78	—	—	—	88	97
79	—	—	—	—	103
80	—	—	—	—	111
81	—	—	—	—	119
82	—	—	—	—	127
83	—	—	—	—	133

*Full-load current (A) does not include FLC x 1.15 or 1.25.

FULL-LOAD CURRENTS—DC MOTORS

Motor rating (HP)	Current (A)	
	120 V	240 V
1/4	3.1	1.6
1/3	4.1	2.0
1/2	5.4	2.7
3/4	7.6	3.8
1	9.5	4.7
1 1/2	13.2	6.6
2	17	8.5
3	25	12.2
5	40	20
7 1/2	48	29

FULL-LOAD CURRENTS—1φ, AC MOTORS

Motor rating (HP)	Current (A)	
	115 V	230 V
1/6	4.4	2.2
1/4	5.8	2.9
1/3	7.2	3.6
1/2	9.8	4.9
3/4	18.8	6.9
1	16	8
1 1/2	20	10
2	24	12
3	34	17
5	56	28
7 1/2	80	40
10	100	50

FULL-LOAD CURRENTS—3φ, AC INDUCTION MOTORS

Motor rating (HP)	Current (A)			
	208 V	230 V	460 V	575 V
1/4	1.11	.96	.48	.38
1/3	1.34	1.18	.59	.47
1/2	2.2	2.0	1.0	.8
3/4	3.1	2.8	1.4	1.1
1	4.0	3.6	1.8	1.4
1 1/2	5.7	5.2	2.6	2.1
2	7.5	6.8	3.4	2.7
3	10.6	9.6	4.8	3.9
5	16.7	15.2	7.6	6.1
7 1/2	24.0	22.0	11.0	9.0
10	31.0	28.0	14.0	11.0
15	46.0	42.0	21.0	17.0
20	59	54	27	22
25	75	68	34	27
30	88	80	40	32
40	114	104	52	41
50	143	130	65	52
60	169	154	77	62
75	211	192	96	77
100	273	248	124	99
125	343	312	156	125
150	396	360	180	144
200	—	480	240	192
250	—	602	301	242
300	—	—	362	288
350	—	—	413	337
400	—	—	477	382
500	—	—	590	472

TYPICAL MOTOR EFFICIENCIES

HP	Standard Motor (%)	Energy-Efficient Motor (%)	HP	Standard Motor (%)	Energy-Efficient Motor (%)
1	76.5	84.0	30	88.1	93.1
1.5	78.5	85.5	40	89.3	93.6
2	79.9	86.5	50	90.4	93.7
3	80.8	88.5	75	90.8	95.0
5	83.1	88.6	100	91.6	95.4
7.5	83.8	90.2	125	91.8	95.8
10	85.0	90.3	150	92.3	96.0
15	86.5	91.7	200	93.3	96.1
20	87.5	92.4	250	93.6	96.2
25	88.0	93.0	300	93.8	96.5

CONTROL RATINGS

Size	Load (V)	Maximum HP Normal duty 1φ	Normal duty 3φ	Plugging & jogging duty 1φ	Plugging & jogging duty 3φ	Cont amps	Service limit amps	Tungsten & ballast type lamp amps 480 V max	Resistance heating (kW) 1φ	Resistance heating (kW) 3φ	Transformer switching 20 times 1φ	20 times 3φ	20-40 times 1φ	20-40 times 3φ	Capacitor kVA switching rating 3φ KVAR
00	115	1/2	—	—	—	9	11	—	1.15	2.0	—	—	—	—	—
	200	—	1 1/2	—	—	9	11	—	2.0	3.46	—	—	—	—	—
	230	1	1 1/2	—	—	9	11	—	2.3	4.0	—	—	—	—	—
	380	—	1 1/2	—	—	9	11	—	—	6.5	—	—	—	—	—
	460	—	2	—	—	9	11	—	4.6	8.0	—	—	—	—	—
	575	—	2	—	—	9	11	—	5.8	10.0	—	—	—	—	—
0	115	1	—	1/2	—	18	21	20	2.3	4.0	0.6	—	—	—	—
	200	—	3	—	1 1/2	18	21	20	4.0	6.92	—	—	—	—	—
	230	2	3	1	1 1/2	18	21	20	4.6	8.0	1.2	1.8	0.3	0.9	—
	380	—	5	—	1 1/2	18	21	20	—	13.1	—	2.1	—	1.0	—
	460	—	5	—	2	18	21	—	9.2	15.9	2.4	4.2	1.2	2.1	—
	575	—	5	—	2	18	21	—	11.5	19.9	3.0	5.2	1.5	2.6	—
1	115	2	—	1	—	27	32	30	3.5	6.0	1.2	—	0.6	—	—
	200	—	7 1/2	—	3	27	32	30	6	10.4	—	—	—	—	—
	230	3	7 1/2	2	3	27	32	30	6.9	11.9	2.4	3.6	1.2	1.8	—
	380	—	10	—	5	27	32	30	—	19.7	—	4.3	—	2.1	—
	460	—	10	—	5	27	32	—	13.8	23.9	4.9	8.5	2.5	4.3	—
	575	—	10	—	5	27	32	—	17.3	29.8	6.2	11.0	3.1	5.3	—

CONTROL RATINGS

Size	Load (V)	Maximum HP — Normal duty 1φ	Normal duty 3φ	Plugging & jogging duty 1φ	Plugging & jogging duty 3φ	Cont amps	Service limit amps	Tungsten & ballast type lamp amps 480 V max	Resistance heating (kW) 1φ	Resistance heating (kW) 3φ	Transformer 20 times 1φ	20 times 3φ	20–40 times 1φ	20–40 times 3φ	Capacitor kVA switching ratings 3φ kVAR
1P	115	3	—	1 1/2	—	35	42	45	5.8	—	—	—	—	—	—
	230	5	—	3	—	35	42	45	11.5	—	—	—	—	—	—
1 3/4	115	—	—	—	—	40	40	45	5.8	9.9	1.6	—	0.8	—	—
	200	—	10	—	5	40	40	45	10	17.3	—	4.9	—	2.4	—
	230	—	10	—	5	40	40	45	11.5	19.9	3.2	5.75	1.6	2.8	—
	380	—	15	—	7 1/2	40	40	45	—	32.9	—	—	—	—	—
	460	—	15	—	7 1/2	40	40	45	23	39.8	6.6	11.2	3.3	5.7	—
	575	—	15	—	7 1/2	40	40	—	28.8	49.7	8.1	14.5	4.1	7.1	—
2	115	3	—	2	—	45	52	60	8.1	13.9	2.1	—	1.0	—	—
	200	—	10	—	—	45	52	60	14	24.2	—	6.3	—	3.1	—
	230	7 1/2	15	5	7 1/2	45	52	60	16.1	27.8	4.1	7.2	2.1	3.6	8
	380	—	25	—	10	45	52	60	—	46.0	—	14	—	7.2	—
	460	—	25	—	15	45	52	60	32.2	55.7	8.3	18	4.2	8.9	16
	575	—	25	—	15	45	52	—	40.3	69.6	10.0	—	5.2	—	20
2 1/2	115	5	—	—	—	60	65	75	10.4	17.9	3.1	—	1.5	—	—
	200	—	15	—	10	60	65	75	18	31.1	—	9.1	—	4.6	—
	230	10	20	—	15	60	65	75	20.7	35.8	6.1	10.6	3.1	5.3	17.5
	380	—	30	—	20	60	65	75	—	59.2	—	21	—	10.6	—
	460	—	30	—	20	60	65	75	41.4	71.6	12	26.5	6.1	13.4	34.5
	575	—	30	—	20	60	65	—	51.8	89.5	15	—	7.6	—	43.5

CONTROL RATINGS

Size	Load (V)	Maximum HP Normal duty 1φ	Normal duty 3φ	Plugging & jogging duty 1φ	Plugging & jogging duty 3φ	Cont amps	Service limit amps	Tungsten & ballast type lamp amps 480 V max	Resistance heating (kW) 1φ	Resistance heating (kW) 3φ	Transformer 20 times 1φ	Transformer 20 times 3φ	Transformer 20-40 times 1φ	Transformer 20-40 times 3φ	Capacitor kVA switching ratings 3φ kVAR
3	115	7 1/2	—	—	—	90	104	100	14.4	24.8	4.1	—	2.0	—	—
	200	—	25	—	15	90	104	100	25	43.3	—	12	—	6.1	—
	230	15	30	—	20	90	104	100	28.8	50.0	8.1	14	4.1	7.0	27
	380	—	50	—	30	90	104	100	—	82.2	—	—	—	—	—
	460	—	50	—	30	90	104	100	57.5	99.4	16	28	8.1	14	53
	575	—	50	—	30	90	104	—	71.9	124	20	35	10	18	67
3 1/2	115	—	—	—	—	115	125	150	18.4	31.8	—	—	—	—	—
	200	—	30	—	20	115	125	150	32	55.4	—	16	—	8	—
	230	—	60	—	25	115	125	150	36.8	63.7	11	18.5	5.4	9.5	33.5
	380	—	60	—	30	115	125	150	—	105	—	—	—	—	—
	460	—	75	—	40	115	125	150	73.6	127	21.5	37.5	11.0	18.5	66.5
	575	—	75	—	40	115	125	—	92	159	37	47	13.5	23.5	83.5
4	200	—	40	—	25	135	156	200	39	67.5	—	20	—	10	—
	230	—	50	—	30	135	156	200	44.9	77.6	14	23	6.8	12	40
	380	—	75	—	50	135	156	200	—	128	—	—	—	—	—
	460	—	100	—	60	135	156	200	89.7	155	27	47	14	23	80
	575	—	100	—	60	135	156	—	112	194	34	59	17	29	100

CONTROL RATINGS

Size	Load (V)	Maximum HP Normal duty 1φ	Maximum HP Normal duty 3φ	Plugging & jogging duty 1φ	Plugging & jogging duty 3φ	Cont amps	Service limit amps	Tungsten & ballast type lamp amps 480 V max	Resistance heating (kW) 1φ	Resistance heating (kW) 3φ	Transformer 20 times 1φ	Transformer 20 times 3φ	Transformer 40 times 1φ	Transformer 40 times 3φ	Capacitor kVA switching ratings 3φ kVAR
4 1/2	200	—	50	—	30	210	225	250	53	91.7	—	30.5	—	15	—
	230	—	75	—	40	210	225	250	60.9	105	20.5	35	10.4	18	60
	380	—	100	—	75	210	225	250	—	174	—	—	—	—	—
	460	—	150	—	100	210	225	250	122	211	40.5	70.5	20.5	35	120
	575	—	150	—	100	210	225	—	152	264	51	88	25.5	44	150

(Transformer switching 50—60 Hz kVA rating inrush peak time — Continuous amps; 20 times and 40 times)

STANDARD MOTOR SIZES

Classification	Size (HP)
Milli	1, 1.5, 2, 3, 5, 7.5, 10, 15, 25, 35
Fractional	1/20, 1/12, 1/8, 1/6, 1/4, 1/3, 1/2, 3/4
Full	1, 1-1/2, 2, 3, 5, 7-1/2, 10, 15, 20, 25, 30, 40, 50, 60, 75, 100, 125, 150, 200, 250, 300
Full—Special Order	350, 400, 450, 500, 600, 700, 800, 900, 1000, 1250, 1500, 1750, 2000, 2250, 2500, 3000, 3500, 4000, 4500, 5000, 5500, 6000, 7000, 8000, 9000, 10,000, 11,000, 12,000, 13,000, 14,000, 15,000, 16,000, 17,000, 18,000, 19,000, 20,000, 22,500, 30,000, 32,500, 35,000, 37,500, 40,000, 45,000, 50,000

MOTOR FRAME DIMENSIONS

Frame No.	Shaft U	Shaft V	Key W	Key T	Key L	A	B	D	E	F	BA
48	1/2	1 1/2*	flat	3/64	—	5 5/8*	3 1/2*	3	2 1/8	1 3/8	2 1/2
56	5/8	1 7/8*	3/16	3/16	1 3/8	6 1/2*	4 1/4*	3 1/2	2 7/16	1 1/2	2 3/4
143T	7/8	2	3/16	3/16	1 3/8	7	6	3 1/2	2 3/4	2	2 1/4
145T	7/8	2	3/16	3/16	1 3/8	7	6	3 1/2	2 3/4	2 1/2	2 1/4
182	7/8	2	3/16	3/16	1 3/8	9	6 1/2	4 1/2	3 3/4	2 1/8	2 3/4
182T	1 1/8	2 1/2	1/4	1/4	1 3/4	9	6 1/2	4 1/2	3 3/4	2 1/8	2 3/4
184	7/8	2	3/16	3/16	1 3/8	9	7 1/2	4 1/2	3 3/4	2 3/4	2 3/4
184T	1 1/8	2 1/2	1/4	1/4	1 3/4	9	7 1/2	4 1/2	3 3/4	2 3/4	2 3/4
203	3/4	2	3/16	3/16	1 3/8	10	7 1/2	5	4	2 3/4	3 1/8
204	3/4	2	3/16	3/16	1 3/8	10	8 1/2	5	4	3 1/4	3 1/8
213	1 1/8	2 3/4	1/4	1/4	2	10 1/2	7 1/2	5 1/4	4 1/4	2 3/4	3 1/2
213T	1 3/8	3 1/4	5/16	5/16	2 3/8	10 1/2	7 1/2	5 1/4	4 1/4	2 3/4	3 1/2
215	1 1/8	2 3/4	1/4	1/4	2	10 1/2	9	5 1/4	4 1/4	3 1/2	3 1/2
215T	1 3/8	3 1/8	5/16	5/16	2 3/8	10 1/2	9	5 1/4	4 1/4	3 1/2	3 1/2
224	1	2 3/4	1/4	1/4	2	11	8 3/4	5 1/2	4 1/2	3 3/8	3 1/2
225	1	2 3/4	1/4	1/4	2	11	9 1/2	5 1/2	4 1/2	3 3/4	3 1/2
254	1 1/8	3 1/8	1/4	1/4	2 3/8	12 1/2	10 3/4	6 1/4	5	4 1/8	4 1/4
254U	1 3/8	3 1/2	5/16	5/16	2 3/4	12 1/2	10 3/4	6 1/4	5	4 1/8	4 1/4
254T	1 5/8	3 3/4	3/8	3/8	2 7/8	12 1/2	10 3/4	6 1/4	5	4 1/8	4 1/4
256U	1 3/8	3 1/2	5/16	5/16	2 3/4	12 1/2	12 1/2	6 1/4	5	5	4 1/4
256T	1 5/8	3 3/4	3/8	3/8	2 7/8	12 1/2	12 1/2	6 1/4	5	5	4 1/4
284	1 1/4	3 1/2	1/4	1/4	2 3/4	14	12 1/2	7	5 1/2	4 3/4	4 3/4
284U	1 5/8	4 5/8	3/8	3/8	3 3/4	14	12 1/2	7	5 1/2	4 3/4	4 3/4
284T	1 7/8	4 3/8	1/2	1/2	3 1/4	14	12 1/2	7	5 1/2	4 3/4	4 3/4
284TS	1 5/8	3	3/8	3/8	1 7/8	14	12 1/2	7	5 1/2	4 3/4	4 3/4
286U	1 5/8	4 5/8	3/8	3/8	3 3/4	14	14	7	5 1/2	5 1/2	4 3/4
286T	1 7/8	4 3/8	1/2	1/2	3 1/4	14	14	7	5 1/2	5 1/2	4 3/4
286TS	1 5/8	3	3/8	3/8	1 7/8	14	14	7	5 1/2	5 1/2	4 3/4
324	1 5/8	4 5/8	3/8	3/8	3 3/4	16	14	8	6 1/4	5 1/4	5 1/4
324U	1 7/8	5 3/8	1/2	1/2	4 1/4	16	14	8	6 1/4	5 1/4	5 1/4
324S	1 5/8	3	3/8	3/8	1 7/8	16	14	8	6 1/4	5 1/4	5 1/4
324T	2 1/8	5	1/2	1/2	3 7/8	16	14	8	6 1/4	5 1/4	5 1/4
324TS	1 7/8	3 1/2	1/2	1/2	2	16	14	8	6 1/4	5 1/4	5 1/4
326	1 5/8	4 5/8	3/8	3/8	3 3/4	16	15 1/2	8	6 1/4	6	5 1/4
326U	1 7/8	5 3/8	1/2	1/2	4 1/4	16	15 1/2	8	6 1/4	6	5 1/4
326S	1 5/8	3	3/8	3/8	1 7/8	16	15 1/2	8	6 1/4	6	5 1/4
326T	2 1/8	5	1/2	1/2	3 7/8	16	15 1/2	8	6 1/4	6	5 1/4
326TS	1 7/8	3 1/2	1/2	1/2	2	16	15 1/2	8	6 1/4	6	5 1/4

*Not NEMA standard dimensions

MOTOR FRAME DIMENSIONS

Frame No	Shaft U	Shaft V	Key W	Key T	Key L	A	B	D	E	F	BA
364	1 7/8	5 3/8	1/2	1/2	4 1/4	18	15 1/4	9	7	5 5/8	5 7/8
364S	1 5/8	3	3/8	3/8	1 7/8	18	15 1/4	9	7	5 5/8	5 7/8
364U	2 1/8	6 1/8	1/2	1/2	5	18	15 1/4	9	7	5 5/8	5 7/8
364US	1 7/8	3 1/2	1/2	1/2	2	18	15 1/4	9	7	5 5/8	5 7/8
364T	2 3/8	5 5/8	5/8	5/8	4 1/4	18	15 1/4	9	7	5 5/8	5 7/8
364TS	1 7/8	3 1/2	1/2	1/2	2	18	15 1/4	9	7	5 5/8	5 7/8
365	1 7/8	5 3/8	1/2	1/2	4 1/4	18	16 1/4	9	7	6 1/8	5 7/8
365S	1 5/8	3	3/8	3/8	1 7/8	18	16 1/4	9	7	6 1/8	5 7/8
365U	2 1/8	6 1/8	1/2	1/2	5	18	16 1/4	9	7	6 1/8	5 7/8
365US	1 7/8	3 1/2	1/2	1/2	2	18	16 1/4	9	7	6 1/8	5 7/8
365T	2 3/8	5 5/8	5/8	5/8	4 1/4	18	16 1/4	9	7	6 1/8	5 7/8
365TS	2 1/8	4	1/2	1/2	2	18	16 1/4	9	7	6 1/8	5 7/8
404	2 1/8	6 1/8	1/2	1/2	5	20	16 1/4	10	8	6 1/8	6 5/8
404S	1 7/8	3 1/2	1/2	1/2	2	20	16 1/4	10	8	6 1/8	6 5/8
404U	2 3/8	6 7/8	5/8	5/8	5 1/2	20	16 1/4	10	8	6 1/8	6 5/8
404US	2 1/8	4	1/2	1/2	2 3/4	20	16 1/4	10	8	6 1/8	6 5/8
404T	2 7/8	7	3/4	3/4	5 5/8	20	16 1/4	10	8	6 1/8	6 5/8
404TS	2 1/8	4	1/2	1/2	2 3/4	20	16 1/4	10	8	6 1/8	6 5/8
405	2 1/8	6 1/8	1/2	1/2	5	20	17 3/4	10	8	6 7/8	6 5/8
405S	1 7/8	3 1/2	1/2	1/2	2	20	17 3/4	10	8	6 7/8	6 5/8
405U	2 3/8	6 7/8	5/8	5/8	5 1/2	20	17 3/4	10	8	6 7/8	6 5/8
405US	2 1/8	4	1/2	1/2	2 3/4	20	17 3/4	10	8	6 7/8	6 5/8
405T	2 7/8	7	3/4	3/4	5 5/8	20	17 3/4	10	8	6 7/8	6 5/8
405TS	2 1/8	4	1/2	1/2	2 3/4	20	17 3/4	10	8	6 7/8	6 5/8
444	2 3/8	6 7/8	5/8	5/8	5 1/2	22	18 1/2	11	9	7 1/4	7 1/2
444S	2 1/8	4	1/2	1/2	2 3/4	22	18 1/2	11	9	7 1/4	7 1/2
444U	2 7/8	8 3/8	3/4	3/4	7	22	18 1/2	11	9	7 1/4	7 1/2
444US	2 1/8	4	1/2	1/2	2 3/4	22	18 1/2	11	9	7 1/4	7 1/2
444T	3 3/8	8 1/4	7/8	7/8	6 7/8	22	18 1/2	11	9	7 1/4	7 1/2
444TS	2 3/8	4 1/2	5/8	5/8	3	22	18 1/2	11	9	7 1/4	7 1/2
445	2 3/8	6 7/8	5/8	5/8	5 1/2	22	20 1/2	11	9	8 1/4	7 1/2
445S	2 1/8	4	1/2	1/2	2 3/4	22	20 1/2	11	9	8 1/4	7 1/2
445U	2 7/8	8 3/8	3/4	3/4	7	22	20 1/2	11	9	8 1/4	7 1/2
445US	2 1/8	4	1/2	1/2	2 3/4	22	20 1/2	11	9	8 1/4	7 1/2
445T	3 3/8	8 1/4	7/8	7/8	6 7/8	22	20 1/2	11	9	8 1/4	7 1/2
445TS	2 3/8	4 1/2	5/8	5/8	3	22	20 1/2	11	9	8 1/4	7 1/2
504U	2 7/8	8 3/8	3/4	3/4	7 1/4	25	21	12 1/2	10	8	8 1/2
504S	2 1/8	4	1/2	1/2	2 3/4	25	21	12 1/2	10	8	8 1/2
505	2 7/8	8 3/8	3/4	3/4	7 1/4	25	23	12 1/2	10	9	8 1/2
505S	2 1/8	4	1/2	1/2	2 3/4	25	23	12 1/2	10	9	8 1/2

COUPLING SELECTIONS

Coupling number	Rated torque (lb-in)	Maximum shock torque (lb-in)
10-101-A	16	45
10-102-A	36	100
10-103-A	80	220
10-104-A	132	360
10-105-A	176	480
10-106-A	240	660
10-107-A	325	900
10-108-A	525	1450
10-109-A	875	2450
10-110-A	1250	3500
10-111-A	1800	5040
10-112-A	2200	6160

TYPICAL MOTOR POWER FACTORS

HP	Speed (rpm)	Power Factor at		
		1/2 load	3/4 load	full load
0—5	1800	.72	.82	.84
5.01—20	1800	.74	.84	.86
20.1—100	1800	.79	.86	.89
100.1—300	1800	.81	.88	.91

V-BELTS

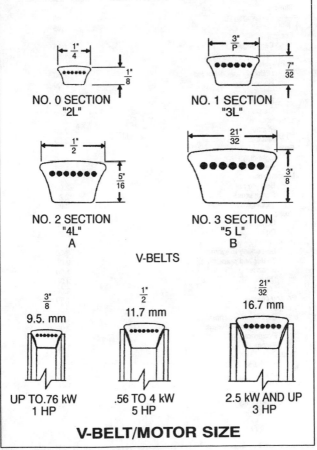

$\frac{1"}{4}$

$\frac{1"}{8}$

NO. 0 SECTION
"2L"

$\frac{3"}{P}$

$\frac{7"}{32}$

NO. 1 SECTION
"3L"

$\frac{1"}{2}$

$\frac{5"}{16}$

NO. 2 SECTION
"4L"
A

$\frac{21"}{32}$

$\frac{3"}{8}$

NO. 3 SECTION
"5 L"
B

V-BELTS

$\frac{3"}{8}$
9.5. mm

$\frac{1"}{2}$
11.7 mm

$\frac{21"}{32}$
16.7 mm

UP TO.76 kW
1 HP

.56 TO 4 kW
5 HP

2.5 kW AND UP
3 HP

V-BELT/MOTOR SIZE

COMMON SERVICE FACTORS

Equipment	Service factor
Blowers	
Centrifugal	1.00
Vane	1.25
Compressors	
Centrifugal	1.25
Vane	1.50
Conveyors	
Uniformly loaded or fed	1.50
Heavy-duty	2.00
Elevators	
Bucket	2.00
Freight	2.25
Extruders	
Plastic	2.00
Metal	2.50
Fans	
Light-duty	1.00
Centrifugal	1.50
Machine tools	
Bending roll	2.00
Punch press	2.25
Tapping machine	3.00
Mixers	
Concrete	2.00
Drum	2.25
Paper mills	
De-barking machines	3.00
Beater and pulper	2.00
Bleacher	1.00
Dryers	2.00
Log haul	2.00
Printing presses	1.50
Pumps	
Centrifugal—general	1.00
Centrifugal—sewage	2.00
Reciprocating	2.00
Rotary	1.50
Textile	
Batchers	1.50
Dryers	1.50
Looms	1.75
Spinners	1.50
Woodworking machines	1.00

MOTOR FRAME TABLE

Frame No. Series	D	1	2	3	4	5	6	7
				Third/Fourth Digit of Frame No.				
140	3.50	3.00	3.50	4.00	4.50	5.00	5.50	6.25
160	4.00	3.50	4.00	4.50	5.00	5.50	6.25	7.00
180	4.50	4.00	4.50	5.00	5.50	6.25	7.00	8.00
200	5.00	4.50	5.00	5.50	6.50	7.00	8.00	9.00
210	5.25	4.50	5.00	5.50	6.25	7.00	8.00	9.00
220	5.50	5.00	5.50	6.25	6.75	7.50	9.00	10.00
250	6.25	5.50	6.25	7.00	8.25	9.00	10.00	11.00
280	7.00	6.25	7.00	8.00	9.50	10.00	11.00	12.50
320	8.00	7.00	8.00	9.00	10.50	11.00	12.00	14.00
360	9.00	8.00	9.00	10.00	11.25	12.25	14.00	16.00
400	10.00	9.00	10.00	11.00	12.25	13.75	16.00	18.00
440	11.00	10.00	11.00	12.50	14.50	16.50	18.00	20.00
500	12.50	11.00	12.50	14.00	16.00	18.00	20.00	22.00
580	14.50	12.50	14.00	16.00	18.00	20.00	22.00	25.00
680	17.00	16.00	18.00	20.00	22.00	25.00	28.00	32.00

MOTOR FRAME TABLE

Frame No. Series	D	Third/Fourth Digit of Frame No.							
		8	9	10	11	12	13	14	15
140	3.50	7.00	8.00	9.00	10.00	11.00	12.50	14.00	16.00
160	4.00	8.00	9.00	10.00	11.00	12.50	14.00	16.00	18.00
180	4.50	9.00	10.00	11.00	12.50	14.00	16.00	18.00	20.00
200	5.00	10.00	11.00	–	–	–	–	–	–
210	5.25	10.00	11.00	12.50	14.00	16.00	18.00	20.00	22.00
220	5.50	11.00	12.50	–	–	–	–	–	–
250	6.25	12.50	14.00	16.00	18.00	20.00	22.00	25.00	28.00
280	7.00	14.00	16.00	18.00	20.00	22.00	25.00	28.00	32.00
320	8.00	16.00	18.00	20.00	22.00	25.00	28.00	32.00	36.00
360	9.00	18.00	20.00	22.00	25.00	28.00	32.00	36.00	40.00
400	10.00	20.00	22.00	25.00	28.00	32.00	36.00	40.00	45.00
440	11.00	22.00	25.00	28.00	32.00	36.00	40.00	45.00	50.00
500	12.50	25.00	28.00	32.00	36.00	40.00	45.00	50.00	56.00
580	14.50	28.00	32.00	36.00	40.00	45.00	50.00	56.00	63.00
680	17.00	36.00	40.00	45.00	50.00	56.00	63.00	71.00	80.00

MOTOR FRAME LETTERS

LETTER	DESIGNATION
G	Gasoline pump motor
K	Sump pump motor
M and N	Oil burner motor
S	Standard short shaft for direct connection
T	Standard dimensions established
U	Previously used as frame designation for which standard dimensions are established
Y	Special mounting dimensions required from manufacturer
Z	Standard mounting dimensions except shaft extension

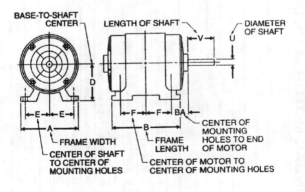

BASE-TO-SHAFT CENTER

LENGTH OF SHAFT

DIAMETER OF SHAFT

V

U

D

E   E

F   F   BA

A

B

FRAME WIDTH

FRAME LENGTH

CENTER OF MOUNTING HOLES TO END OF MOTOR

CENTER OF SHAFT TO CENTER OF MOUNTING HOLES

CENTER OF MOTOR TO CENTER OF MOUNTING HOLES

ANNUAL MOTOR MAINTENANCE CHECKLIST

Motor File #: _____ Serial #: _____
Date Installed: _____ Motor Location: _____

MFR: _____ Type: _____ Frame: _____
HP: _____ Volts: _____ Amps: _____
RPM: _____ Date Serviced: _____

Step	Operation	Mechanic
1	Turn OFF and lock out all power to the motor and its control circuit.	
2	Clean motor exterior and all ventilation ducts.	
3	Uncouple motor from load and disassemble.	
4	Clean inside of motor.	
5	Check centrifugal switch assemblies.	
6	Check rotors, armatures, and field windings	
7	Check all peripheral equipment.	
8	Check bearings.	
9	Check brushes and commutator.	
10	Check slip rings.	
11	Reassemble motor and couple to load.	

ANNUAL MOTOR MAINTENANCE CHECKLIST

Step	Operation	Mechanic
12	Flush old bearing lubricant and replace.	
13	Check motor's wire raceway.	
14	Check drive mechanism.	
15	Check motor terminations.	
16	Check capacitors.	
17	Check all mounting bolts.	
18	Check and record line-to-line resistance.	
19	Check and record megohmmeter resistance from T1 to ground.	
20	Check and record insulation polarization index.	
21	Check motor controls.	
22	Reconnect motor and control circuit power supplies.	
23	Check line-to-line voltage for balance and level.	
24	Check line current draw against nameplate rating	
25	Check and record inboard and outboard bearing temperatures	

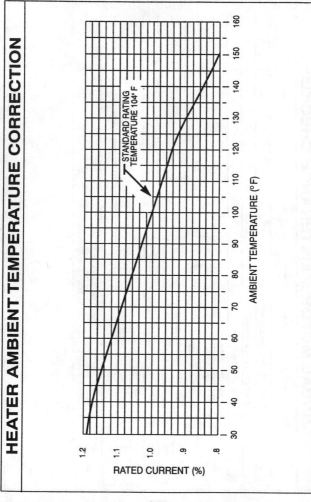

HEATER AMBIENT TEMPERATURE CORRECTION

STANDARD RATING TEMPERATURE 104° F

RATED CURRENT (%)

AMBIENT TEMPERATURE (°F)

MOTOR REPAIR AND SERVICE RECORD

Motor File #: _____   Serial #: _____
Date Installed: _____   Motor Location: _____

MFR: _____   Type: _____   Frame: _____
HP: _____   Volts: _____   Amps: _____
RPM: _____   Filter Sizes: _____

Date	Operation	Mechanic

SEMIANNUAL MOTOR MAINTENANCE CHECKLIST

Motor File #: _____   Serial #: _____
Date Installed: _____   Motor Location: _____

MFR: _____   Type: _____   Frame: _____
HP: _____   Volts: _____   Amps: _____
RPM: _____   Date Serviced: _____

SEMIANNUAL MOTOR MAINTENANCE CHECKLIST

Step	Operation	Mechanic
1	Turn OFF and lock out all power to the motor and its control circuit.	
2	Clean motor exterior and all ventilation ducts.	
3	Check motor's wire raceway.	
4	Check and lubricate bearings as needed.	
5	Check drive mechanism.	
6	Check brushes and commutator.	
7	Check slip rings.	
8	Check motor terminations.	
9	Check capacitors.	
10	Check all mounting bolts.	
11	Check and record line-to-line resistance	
12	Check and record megohmmeter resistance from L1 to ground.	
13	Check motor controls.	
14	Reconnect motor and control circuit power supplies.	
15	Check line-to-line voltage for balance and level.	
16	Check line current draw against nameplate rating.	
17	Check and record inboard and outboard bearing temperatures.	

LOCKED ROTOR CURRENT

Apparent, 1φ	Apparent, 3φ	True, 1φ	True, 3φ
$$LRC = \frac{1000 \times HP \times kVA/HP}{V}$$ where LRC = locked rotor current (in amps) 1000 = multiplier for kilo HP = horsepower kVA/HP = kilovolt amps per horsepower V = volts	$$LRC = \frac{1000 \times HP \times kVA/HP}{V \times \sqrt{3}}$$ where LRC = locked rotor current (in amps) 1000 = multiplier for kilo HP = horsepower kVA/HP = kilovolt amps per horsepower V = volts $\sqrt{3}$ = 1.73	$$LRC = \frac{1000 \times HP \times kVA/HP}{V \times PF \times E_{ff}}$$ where LRC = locked rotor current (in amps) 1000 = multiplier for kilo HP = horsepower kVA/HP = kilovolt amps per horsepower V = volts PF = power factor E_{ff} = motor efficiency	$$LRC = \frac{1000 \times HP \times kVA/HP}{V \times \sqrt{3} \times PF \times E_{ff}}$$ where LRC = locked rotor current (in amps) 1000 = multiplier for kilo HP = horsepower kVA/HP = kilovolt amps per horsepower V = volts $\sqrt{3}$ = 1.73

MAXIMUM OCPD

$OCPD = FLC \times R_M$
where
FLC = full load current (from motor nameplate or NEC® Table 430-150)
R_M = maximum rating of OCPD

Motor Type	Code Letter	FLC %				
		Motor Size	TDF	NTDF	ITB	ITCB
AC*	—	—	175	300	150	700
AC*	A	—	150	150	150	700
AC*	B—E	—	175	250	200	700
AC*	F—V	—	175	300	250	700
DC	—	1/8 TO 50 HP	150	150	150	250
DC	—	Over 50 HP	150	150	150	175

* full-voltage and resistor starting

EFFICIENCY

Input and Output Power Known	Horsepower and Power Loss Known
$E_{ff} = \dfrac{P_{out}}{P_{in}}$	$E_{ff} = \dfrac{746 \times HP}{746 \times HP + W_l}$
where	where
E_{ff} = efficiency (%)	E_{ff} = efficiency (%)
P_{out} = output power (W)	746 = constant
P_{in} = input power (W)	HP = horsepower
	W_l = watts lost

VOLTAGE UNBALANCE

$$V_u = \frac{V_d}{V_a} \times 100$$

where
V_u = voltage unbalance (%)
V_d = voltage deviation (V)
V_a = voltage average (V)
100 = constant

POWER

Power Consumed	Operating Cost	Annual Savings
$P = \dfrac{HP \times 746}{E_{ff}}$ where P = power consumed (W) HP = horsepower 746 = constant E_{ff} = efficiency (%)	$C_{/hr} = \dfrac{P_{/hr} \times C_{/kWh}}{1000}$ where $C_{/hr}$ = operating cost per hour $P_{/hr}$ = power consumed per hour $C_{/kWh}$ = cost per kilowatt hour 1000 = constant to remove kilo	$S_{Ann} = C_{Ann\ Std} - C_{Ann\ Eff}$ S_{Ann} = annual cost savings $C_{Ann\ Std}$ = annual operating cost for standard motor $C_{Ann\ Eff}$ = annual operating cost for energy-efficient motor

6-78

FAULTY SOLENOID PROBLEMS

Problem	Possible Causes	Comments
Failure to operate when energized	Complete loss of power to solenoid	Normally caused by blown fuse or control circuit problem.
	Low voltage applied to the solenoid	Voltage should be at least 85% of solenoid's rated value.
	Burned out solenoid coil	Normally evident by pungent odor caused by burnt insulation.
	Shorted coil	Normally a fuse is blown and continues to blow when changed.
	Obstruction of plunger movement	Normally caused by a broken part, misalignment, or the presence of a foreign object.
	Excessive pressure on solenoid plunger	Normally caused by excessive system pressure in solenoid-operated valves.
Failure to operate spring-return solenoids when de-energized	Faulty control circuit	Normally a problem of the control circuit not disengaging the solenoid's hold or memory circuit.
	Obstruction of plunger movement	Normally caused by a broken part, misalignment, or the presence of a foreign object.
	Excessive pressure on solenoid plunger	Normally caused by excessive system pressure in solenoid-operated valves
Failure to operate electrically-operated return solenoids when de-energized	Complete loss of power to solenoid	Normally caused by a blown fuse or control circuit problem.
	Low voltage applied to solenoid	Voltage should be at least 85% of solenoid's rated value.

FAULTY SOLENOID PROBLEMS

Problem	Possible Causes	Comments
Failure to operate electrically operated return solenoids when de-energized	Burned out solenoid coil	Normally evident by pungent odor caused by burnt insulation.
	Obstruction of plunger movement	Normally caused by broken part, misalignment, or presence of a foreign object
	Excessive pressure on solenoid plunger	Normally caused by excessive system pressure in solenoid-operated valves
Noisy operation	Solenoid housing vibrates	Normally caused by loose mounting screws.
	Plunger pole pieces do not make flush contact	An air gap may be present causing the plunger to vibrate. These symptoms are normally caused by foreign matter.
Erratic operation	Low voltage applied to the solenoid	Voltage should be at least 85% of the solenoid's rated voltage.
	System pressure may be low or excessive	Solenoid size is inadequate for the application.
	Control circuit is not operating properly	Conditions on the solenoid have increased to the point where the solenoid cannot deliver the required force.

HEATING ELEMENT SPECIFICATIONS

Material State	Heated Material	Outer Cover Material	Maximum Watt Density*	Material Operating Temperature***
	Gasoline	Iron/steel	18-22	300
	Oil (low viscosity)	Steel/copper	20-25	Up to 180
	Oil (medium viscosity)	Steel/copper	12-18	Up to 180
Liquid	Oil (high viscosity)	Steel/copper	5-7	Up to 180
	Vegetable oil (cooking)	Copper	28-32	400
	Water (washroom)	Copper	70-90	140
	Water (process)	Copper	40-50	212
Air/Gas	Still air (ovens)	Steel/stainless steel	28-32 18-22 8-10 2-3	Up to 700 Up to 1000 Up to 1200 Up to 1500
	Air (moving at 10 fps)	Aluminum steel	30-34 23-26 14-16 2-3	Up to 700 Up to 1000 Up to 1200 Up to 1500

HEATING ELEMENT SPECIFICATIONS

Material State	Heated Material	Outer Cover Material	Maximum Watt Density*	Material Operating Temperature**
Solid	Asphalt	Iron/steel	9-12 8-10 6-8 5-6	Up to 200 Up to 300 Up to 400 Up to 500
	Molten tin (heated in pot)	Iron/steel	20-22	600
	Steel tubing (heated indirectly)	Iron/steel	45-48 50-52 54-56	500 750 1000

VOLTAGE VARIATION CHARACTERISTICS

Performance Characteristics	10% above Rated Voltage	10% below Rated Voltage
Starting current	+10% to +12%	−10% to −12%
Full-load current	-7%	+11%
Motor torque	+20% to +25%	−20% to −25%
Motor efficiency	Little change	Little change
Speed	+1%	-1.5%
Temperature rise	−3°C to −4°C	+6°C to +7°C

FREQUENCY VARIATION CHARACTERISTICS

Performance Characteristics	5% above Rated Frequency	5% below Rated Frequency
Starting current	−5% to −6%	+5% to +6%
Full-load current	−1%	+1%
Motor torque	−10%	+11%
Motor efficiency	Slight increase	Slight decrease
Speed	+5%	−5%
Temperature rise	Slight decrease	Slight increase

DC MOTOR PERFORMANCE CHARACTERISTICS

Performance Characteristics	Voltage 10% below Rated Voltage		Voltage 10% above Rated Voltage	
	Shunt	Compound	Shunt	Compound
Starting Torque	−15%	−15%	+15%	+15%
Speed	−5%	−6%	+5%	+6%
Current	+12%	+12%	−8%	−8%
Field temperature	Increases	Decreases	Increases	Increases
Armature temperature	Increases	Increases	Decreases	Decreases
Commutator temperature	Increases	Increases	Decreases	Decreases

MAXIMUM ACCELERATION TIME

Frame Number	Maximum Acceleration Time (in seconds)
48 and 56	8
143-286	10
324-326	12
364-505	15

CHAPTER 7
TRANSFORMERS

TRANSFORMER CONNECTIONS

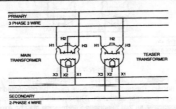

SCOTT CONNECTED THREE-PHASE TO TWO-PHASE

In some localities, two-phase power is required from a three-phase system. The Scott connection is the most popular method of making this phase change. The secondary may be either three, four, or five wire. Special taps must be provided at 50% and 86.6% of normal primary voltage in order to make this connection.

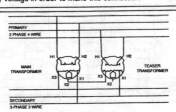

SCOTT CONNECTED TWO-PHASE TO THREE-PHASE

If it should be necessary to supply three-phase power from a two-phase system, the Scott connection may again be used. In this case, of course, the special taps must be provided on the secondary side. In other respects, the connection is similar to the three-phase to two-phase transformation.

If it is desired to obtain the Scott transformation without a special 86.6% tapped transformer, it is possible to use one with 10% or two 5% taps to approximate the desired value. It will introduce a small error of unbalance (overvoltage) which will require care in application.

TRANSFORMER CONNECTIONS

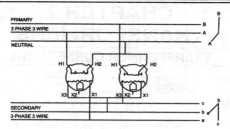

OPEN Y-DELTA

When operating Y-delta and one phase is disabled, service may be maintained at reduced load as shown. The neutral in this case must be connected to the neutral of the setup bank through a copper conductor. The system is unbalanced, electro-statically and electro-magnetically, so that telephone interference may be expected if the neutral is connected to ground. The useful capacity of the open delta open Y bank is 87% of the capacity of the installed transformers when the two units are identical.

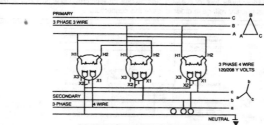

DELTA-Y FOR LIGHTING AND POWER

In the previous banks the single-phase lighting load is all on one phase resulting in unbalanced primary currents in any one bank. To eliminate this difficulty, the delta-Y system finds many uses. Here the neutral of the secondary three-phase system is grounded and the single-phase loads are connected between the different phase wires and the neutral while the three-phase loads are connected to the phase wires. Thus, the single-phase load can be balanced on three phases in each bank and banks may be paralleled if desired.

TRANSFORMER CONNECTIONS

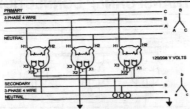

Y-Y FOR LIGHTING AND POWER

This diagram shows a system on which the primary voltage was increased from 2400v to 4160v to increase the potential capacity of the system. The previously delta connected distribution transformers are now connected from line to neutral. The secondaries are connected in Y. In this system the primary neutral is connected to the neutral of the supply voltage through a metallic conductor and carried with the phase conductor to minimize telephone interference. If the neutral of the transformer is isolated from the system neutral as unstable condition results at the transformer neutral caused primarily by third harmonic voltages. If the transformer neutral is connected to ground, the possibility of telephone interference is greatly enhanced and there is also a possibility of resonance between the line capacitance to ground and the magnetizing impedance of the transformer.

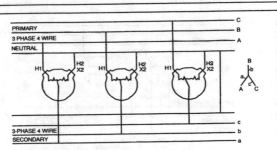

Y-Y AUTOTRANSFORMERS FOR SUPPLYING POWER
FROM A THREE-PHASE FOUR-WIRE SYSTEM

When the ratio of transformation from the primary to secondary voltage is small, the most economical way of stepping down voltage is by using autotransformers as shown. For this application, it is necessary that the neutral of the autotransformer bank be connected to the system neutral.

TRANSFORMER CONNECTIONS

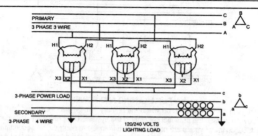

DELTA-DELTA FOR POWER AND LIGHTING

This connection is often used to supply a small single-phase lighting load and three-phase power load simultaneously. As shown in diagram, the midtap of the secondary of one transformer is grounded. Thus, the small lighting load is connected across the transformer with the mid-tap and the ground wire common to both 120v circuits. The single-phase lighting load reduces the available three-phase capacity. This connection requires special watt-hour metering.

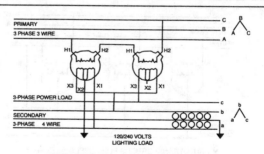

OPEN-DELTA FOR LIGHTING AND POWER

Where the secondary load is a combination of lighting and power the open-delta connected bank is frequently used. This connection is used when the single-phase lighting load is large as compared with the power load. Here two different size transformers may be used with the lighting load connected across the larger rated unit.

TRANSFORMER CONNECTIONS

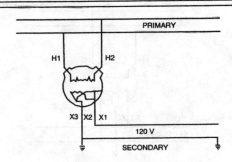

SINGLE-PHASE TO SUPPLY 120V LIGHTING LOAD

The transformer is connected between high voltage line and load with the 120/240v winding connected in parallel. This connection is used where the load is comparatively small and the length of the secondary circuit is short. It is often used for a single customer.

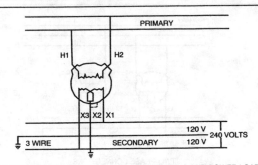

SINGLE-PHASE TO SUPPLY 120/240V 3-WIRE LIGHTING AND POWER LOAD

Here the 120/240v winding is connected in series and the mid-point brought out, making it possible to serve both 120 and 240v loads simultaneously. This connection is used in most urban distribution circuits.

TRANSFORMER CONNECTIONS

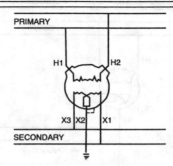

PRIMARY

H1   H2

X3 X2 X1

SECONDARY

SINGLE-PHASE FOR POWER

In this case the 120/240v winding is connected in series serving 240v on a two-wire system. This connection is used for small industrial applications.

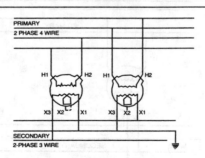

PRIMARY
2 PHASE 4 WIRE

H1   H2   H1   H2

X3 X2 X1   X3 X2 X1

SECONDARY
2-PHASE 3 WIRE

TWO-PHASE CONNECTIONS

This connection consists merely of two single-phase transformers operated 90° out of phase. For a three-wire secondary as shown, the common wire must carry **32 times the load current. In some cases, a four-wire or a five-wire secondary may be used.

AC MOTORS

SINGLE-PHASE

FULL LOAD AMPERES

HP	115 V	208 V	230 V	MIN. TRANSFORMER kVA
1/6	4.4	2.4	2.2	.53
1/4	5.8	3.2	2.9	.70
1/3	7.2	4.0	3.6	.87
1/2	9.8	5.4	4.9	1.18
3/4	13.8	7.6	6.9	1.66
1	16	8.8	8	1.92
1-1/2	20	11	10	2.4
2	24	13.2	12	2.88
3	34	18.7	17	4.1
5	56	30.8	28	6.72
7-1/2	80	44	40	9.6
10	100	55	50	12

AC MOTORS

THREE-PHASE

FULL LOAD AMPERES

HP	208 V	230 V	460 V	575 V	MIN. TRANSFORMER kVA
1/2	2.2	2.0	1.0	0.8	0.9
3/4	3.1	2.8	1.4	1.1	1.2
1	4.0	3.6	1.8	1.4	1.5
1-1/2	5.7	5.2	2.6	2.1	2.1
2	7.5	6.8	3.4	2.7	2.7
3	10.6	9.6	4.8	3.9	3.8
5	16.7	15.2	7.6	6.1	6.3
7-1/2	24.2	22	11	9	9.2
10	31.8	28	14	11	11.2
15	46.2	42	21	17	16.6
20	59.4	54	27	22	21.6
25	74.8	68	34	27	26.6
30	88	80	40	32	32.4
40	114.4	104	52	41	43.2
50	143	130	65	52	52
60	169.4	154	77	62	64
75	211.2	192	96	77	80
100	272.8	248	124	99	103
125	343.2	312	156	125	130
150	396	360	180	144	150
200	528	480	240	192	200

TRANSFORMERS

SINGLE-PHASE

kVA RATING	AMPERES			
	120 V	240 V	480 V	600 V
1	8.33	4.17	2.08	1.67
1-1/2	12.5	6.25	3.13	2.50
2	16.7	8.33	4.17	3.33
3	25.0	12.5	6.25	5.00
5	41.7	20.8	10.4	8.33
7-1/2	62.5	31.3	15.6	12.5
10	83.3	41.7	20.8	16.7
15	125	62.5	31.3	25.0
20	167	83.3	41.7	33.3
25	208	104	52.1	41.7
30	250	125	62.5	50
37-1/2	313	156	78.0	62.5
50	417	208	104	83.3
75	625	313	156	125
100	833	417	208	167
150	1250	625	313	250
167	1392	696	348	278
200	1667	833	417	333
250	2083	1042	521	417
333	2775	1388	694	555
500	4167	2083	1042	833

TRANSFORMERS

THREE-PHASE

kVA RATING	AMPERES			
	208 V	240 V	480 V	600 V
3	8.3	7.2	3.6	2.9
6	16.6	14.4	7.2	5.8
9	25.0	21.6	10.8	8.7
15	41.6	36	18	14.4
20	55.6	48.2	24.1	19.3
25	69.5	60.2	30.1	24.1
30	83.0	72	36	28.8
37-1/2	104	90.3	45.2	36.1
45	125	108	54	43
50	139	120	60.2	48.2
60	167	145	72.3	57.8
75	208	180	90	72
100	278	241	120	96.3
112.5	312	270	135	108
150	415	360	180	144
200	554	480	240	192
225	625	540	270	216
300	830	720	360	288
400	1110	960	480	384
500	1380	1200	600	480
750	2080	1800	900	720
1000	2780	2400	1200	960
1500	4150	3600	1800	1440
2000	5540	4800	2400	1920

RF COIL WINDING DATA

The inductance (I), in microhenrys, of air-core coil can be calculated to within 1% or 2% with the following formulas:

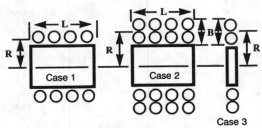

CASE 1: Single Layer Coil

$$I = \frac{R^2 N^2}{9R + 10L}$$

CASE 2: Multiple Layer Coil

$$I = \frac{0.8 \, (R^2 N^2)}{6R + 9L + 10B}$$

CASE 3: Single Layer, Single Row Coil

$$I = \frac{R^2 N^2}{8R + 11B}$$

In all of the above equations, N = number of turns and I is the inductance in microhenrys, L, B and R are measured in inches.

WIRE SIZE vs TURNS/INCH

Gauge AWG	Number of Turns Per Inch of Length		
	Enamel	S.S.C.	D.C.C.
1	-	-	3.3
2	-	-	3.6
3	-	-	4.0
4	-	-	4.5
5	-	-	5.0
6	-	-	5.6
7	-	-	6.2
8	7.6	-	7.1
9	8.6	-	7.8
10	9.6	-	8.9
11	10.7	-	9.8
12	12.0	-	10.9
13	13.5	-	12.0
14	15.0	-	13.8
15	16.8	-	14.7
16	18.9	18.9	16.4
17	21.2	21.2	18.1
18	23.6	23.6	19.8
19	26.4	26.4	21.8
20	29.4	29.4	23.8
21	33.1	32.7	26.0
22	37.0	36.5	30.0
23	41.3	40.6	31.6
24	46.3	45.3	35.6
25	51.7	50.4	38.6
26	58.0	55.6	41.8
27	64.9	61.5	45.0
28	72.7	68.6	48.5
29	81.6	74.8	51.8
30	90.5	83.3	55.5
31	101.0	92.0	59.2
32	113.0	101.0	62.6
33	127.0	110.0	66.3
34	143.0	120.0	70.0
35	158.0	132.0	73.5
36	175.0	143.0	77.0
37	198.0	154.0	80.3
38	224.0	166.0	83.6
39	248.0	181.0	86.6
40	282.0	194.0	89.7

The above values will vary slightly dependiing on the manufacturer of the wire and thickness of enamel.

CHAPTER 8
PLAN SYMBOLS

To avoid confusion, ASA policy requires that the same symbol not be included in more than one Standard. If the same symbol were to be used in two or more Standards, and one of these Standards were revised, changing the meaning of the symbol, considerable confusion could arise over which symbol was correct, the revised or unrevised.

The symbols in this category include, but are not limited to, those listed below.

(MOT)	Electric motor
(GEN)	Electric generator
	Power transformer
	Pothead (cable termination)
(WH)	Electric watthour meter
CB	Circuit element, e.g., circuit breaker
	Circuit breaker
	Fusible element
	Single-throw knife switch

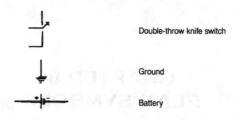

Double-throw knife switch

Ground

Battery

| Ceiling | Wall | **LIGHTING OUTLETS** |

Surface or pendant incandescent, mercury vapor, or similar lamp fixture

Recessed incandescent, mercury vapor, or similar lamp fixture

Surface or pendant individual fluorescent fixture

Recessed individual fluorescent fixture

Surface or pendant continuous-row fluorescent fixture

Recessed continuous-row fluorescent fixture*

Bare-lamp fluorescent strip**

---

* In the case of combination continuous-row fluorescent and incandescent spotlights, use combinations of the above Standard symbols.
** In the case of a continuous-row bare-lamp fluorescent strip above an area-wide diffusion means, show each fixture run, using the Standard symbol; indicate area of diffusing means and type of light shading and/or drawing notation.

$(\dot{X})$ $—(X)$ Surface or pendant exit light

$(\overline{XR})$ $—(XR)$ Recessed exit light

(B) $—(B)$ Blanked outlet

(J) $—(J)$ Junction box

(L) $—(L)$ Outlet controlled by low-voltage switching when relay is installed in outlet box

RECEPTACLE OUTLETS

Unless noted to the contrary, it should be assumed that every receptacle will be grounded, and will have a separate grounding contact.

Single receptacle outlet

Duplex receptacle outlet

Triplex receptacle outlet

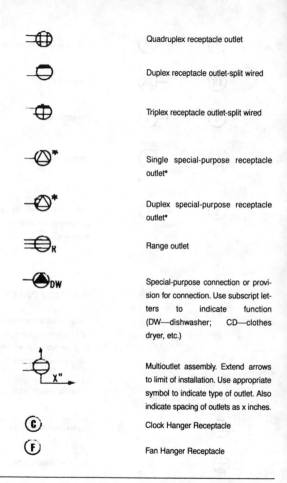

Quadruplex receptacle outlet

Duplex receptacle outlet-split wired

Triplex receptacle outlet-split wired

Single special-purpose receptacle outlet*

Duplex special-purpose receptacle outlet*

Range outlet

Special-purpose connection or provision for connection. Use subscript letters to indicate function (DW—dishwasher; CD—clothes dryer, etc.)

Multioutlet assembly. Extend arrows to limit of installation. Use appropriate symbol to indicate type of outlet. Also indicate spacing of outlets as x inches.

Clock Hanger Receptacle

Fan Hanger Receptacle

---

* Use numeral or letter, either within the symbol or as a subscript alongside the symbol keyed to explanation in the drawing list of symbols, to indicate type of receptacle or usage.

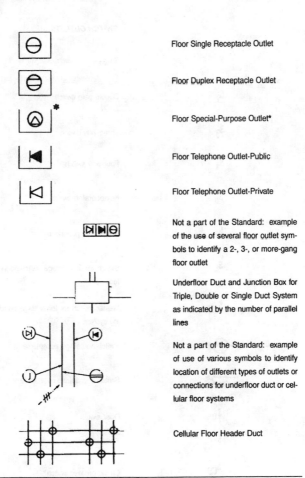

Floor Single Receptacle Outlet

Floor Duplex Receptacle Outlet

Floor Special-Purpose Outlet*

Floor Telephone Outlet-Public

Floor Telephone Outlet-Private

Not a part of the Standard: example of the use of several floor outlet symbols to identify a 2-, 3-, or more-gang floor outlet

Underfloor Duct and Junction Box for Triple, Double or Single Duct System as indicated by the number of parallel lines

Not a part of the Standard: example of use of various symbols to identify location of different types of outlets or connections for underfloor duct or cellular floor systems

Cellular Floor Header Duct

---

*Use numeral or letter, either within the symbol or as a subscript alongside the symbol keyed to explanation in the drawing list of symbols, to indicate type of receptacle or usage.

S	Single-pole switch
S_2	Double-pole switch
S_3	Three-way switch
S_4	Four-way switch
S_K	Key-operated switch
S_P	Switch and pilot lamp
S_L	Switch for low-voltage switching system
S_{LM}	Master switch for low-voltage switching system
⊖S	Switch and single receptacle
⊜S	Switch and double receptacle
S_D	Door switch
S-	Time switch
S_{CB}	Circuit-breaker switch
S_{MC}	Momentary contact switch or push-button for other than signaling system

INSTITUTIONAL, COMMERCIAL, AND INDUSTRIAL OCCUPANCIES

These symbols are recommended by the American Standards Association, but are not used universally. The reader should remember not to assume that these symbols will be used on any certain plan, but to always check the symbol list on the plans, and verify if these symbols are actually used.

Basic Symbol	Examples of Individual Item Identification (Not part of the standard)	

Nurse Call System Devices (any type)

Nurses' Annunciator (can add a number after it as to indicate number of lamps)

Call station, single cord, pilot light

Call station, double cord, microphone speaker

Corridor dome light, 1 lamp

Transformer

Any other item on same system-use numbers as required.

Paging System Devices (any type)

Keyboard

Flush annunciator

2-face annunciator

Any other item on same system-use numbers as required.

Basic Symbol	Examples of Individual Item Identification (Not part of the standard)	
		Fire Alarm System Devices (any type) including Smoke and Sprinkler Alarm Devices
		Control panel
		Station
		10" Gong
		Pre-signal chime
		Any other item on same system—use numbers as required.
		Staff Register System Devices (any type)
		Phone operators' register
		Entrance register—flush
		Staff room register
		Transformer
		Any other item on same system—use numbers as required.

Basic Symbol	Examples of Individual Item Identification (Not part of the standard)	
+⬡		*Electric Clock System Devices (any type)*
	+⟨1⟩	Master clock
	+⟨2⟩	12" Secondary–flush
	+⟨3⟩	12" Double dial–wall mounted
	+⟨4⟩	18" Skeleton dial
	+⟨5⟩	Any other item on same system–use numbers as required.
+◁		*Public Telephone System Devices*
	+◁1	Switchboard
	+◁2	Desk phone
	+◁3	Any other item on same system–use numbers as required.
+◀		*Private Telephone System Devices (any type)*
	+◀1	Switchboard
	+◀2	Wall phone
	+◀3	Any other item on same system–use numbers as required.

Basic Symbol	Examples of Individual Item Identification (Not part of the standard)	
+⌂		*Watchman System Devices (any type)*
	+1	Central station
	+2	Key station
	+3	Any other item on same system—use numbers as required.
+⊐		*Sound System*
	+1	Amplifier
	+2	Microphone
	+3	Interior speaker
	+4	Exterior speaker
	+5	Any other item on same system—use numbers as required.
+⊙		*Other Signal System Devices*
	+11	Buzzer
	+21	Bell
	+31	Pushbutton
	+41	Annunciator
	+51	Any other item on same system—use numbers as required.

8-10

RESIDENTIAL OCCUPANCIES

When a descriptive symbol list is not employed, use the following signaling system symbols to identify standardized, residential-type, signal-system items on residential drawings. Us the basic symbols with a descriptive symbol list when other signal-system items are to be identified.

Pushbutton

Buzzer

Bell

Combination bell-buzzer

Chime

Annunciator

Electric door opener

Maid's signal plug

Interconnection box

Bell-ringing transformer

Outside telephone

Interconnecting telephone

Radio outlet

Television outlet

8-11

PANELBOARDS, SWITCHBOARDS, AND RELATED EQUIPMENT

Flush-mounted panelboard and cabinet*

Surface-mounted panelboard and cabinet*

Switchboard, power control center, unit substations*–should be drawn to scale

Flush-mounted terminal cabinet.* In small-scale drawings the TC may be indicated alongside the symbol.

Surface-mounted terminal cabinet.* In small-scale drawings the TC may be indicated alongside the symbol.

Pull box (identify in relation to wiring section and sizes)

Motor or other power controller*

Externally-operated disconnection switch*

Combination controller and disconnection means*

---

* Identify by notation or schedule.

BUS DUCTS AND WIREWAYS

| T | | T | | T | Trolley duct*

| B | | B | | B | Busway (service, feeder, or plug-in)*

| C | | C | | C | Cable trough ladder or channel*

| W | | W | | W | Wireway*

REMOTE CONTROL STATIONS FOR MOTORS OR OTHER EQUIPMENT

Pushbutton station

| F | Float switch–mechanical

| L | Limit switch–mechanical

| P | Pneumatic switch–mechanical

Electric eye–beam source

Electric eye–relay

—Ⓣ   Thermostat

* Identify by notation or schedule.

8-13

CIRCUITING

Wiring method identification by notation on drawing or in specification.

————————— Wiring concealed in ceiling or wall

- - - - - - - - - Wiring concealed in floor

– – – – – – – – – Wiring exposed

Note: Use heavyweight line to identify service and feeders. Indicate empty conduit by notation CO (conduit only).

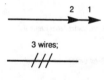

3 wires;

Branch-circuit home run to panelboard. Number of arrows indicates number of circuits. (A numeral at each arrow may be used to identify circuit number.) Note: Any circuit without further identification indicates two-wire circuit. For a greater number of wires, indicate with cross lines, e.g.:

4 wires;

Unless indicated otherwise, the wire size of the circuit is the minimum size required by the specification.

Identify different functions of wiring system, e.g., signaling system by notation or other means.

————————o Wiring turned up

————————● Wiring turned down

8-14

| M | Manhole* |

| H | Handhole* |

| TM | Transformer manhole or vault* |

| TP | Transformer pad* |

Underground direct burial cable. Indicate type, size, and number of conductors by notation or schedule.

Underground duct line. Indicate type, size, and number of ducts by cross-section identification of each run by notation or schedule. Indicate type, size, and number of conductors by notation or schedule.

Streetlight standard feed from underground circuit*

---

* Identify by notation or schedule.

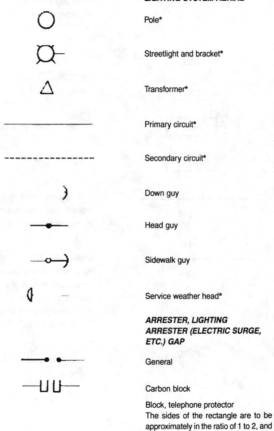

Pole*

Streetlight and bracket*

Transformer*

Primary circuit*

Secondary circuit*

Down guy

Head guy

Sidewalk guy

Service weather head*

ARRESTER, LIGHTING ARRESTER (ELECTRIC SURGE, ETC.) GAP

General

Carbon block

Block, telephone protector
The sides of the rectangle are to be approximately in the ratio of 1 to 2, and the space between rectangles shall be approximately equal to the width of a rectangle.

Identify by notation or schedule.

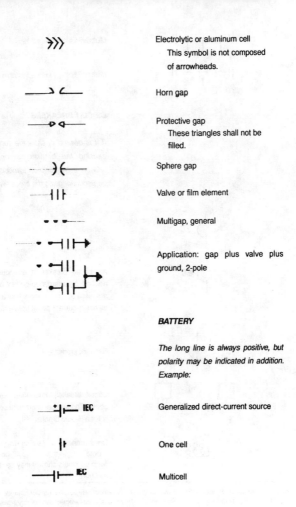

Electrolytic or aluminum cell
   This symbol is not composed
   of arrowheads.

Horn gap

Protective gap
   These triangles shall not be
   filled.

Sphere gap

Valve or film element

Multigap, general

Application: gap plus valve plus
ground, 2-pole

BATTERY

*The long line is always positive, but
polarity may be indicated in addition.
Example:*

Generalized direct-current source

One cell

Multicell

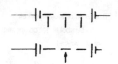 Multicell battery with 3 taps

Multicell battery with adjustable tap

CIRCUIT BREAKERS

If it is desired to show the condition causing the breaker to trip, the relayprotective-function symbols may be used alongside the break symbol.

 IEC   General

IEC   Air circuit breaker, if distinction is needed; for alternating-current breakers rated at 1,500 volts or less and for all direct-current circuit breakers.

Network protector

 IEC   IEC   Circuit breaker, other than covered above. The symbol in the right column is for a 3-pole breaker.

On a connection or wiring diagram, a 3-pole single-throw circuit breaker (with terminals shown) may be drawn as shown.

Note - General Circuit Breaker – On a power diagram, the symbol may be used without other identification. On a composite drawing where confusion with the general circuit element symbol may result, add the identifying letter CB inside or adjacent to the square.

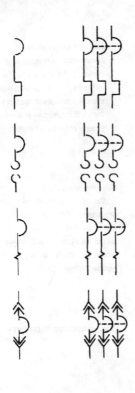

Applications

3-pole circuit breaker with thermal overload device in all 3 poles

3-pole circuit breaker with magnetic overload device in all 3 poles

3-pole circuit breaker, drawout type

CIRCUIT RETURN

Ground

a) A direct conducting connection to the earth or body of water that is a part thereof

b) A conducting connection to a structure that serves a function similar to that of an earth ground (that is, a structure such as a frame of an air, space, or land vehicle that is not conductively connected to earth)

8-19

IEC

Chassis or frame connection

A conducting connection to a chassis or frame of a unit. The chassis or frame may be a sub-stantial potential with respect to the earth or structure in which this chassis or frame is mounted.

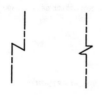
*

Common connections

Conducting connections made to one another. All like-designated points are connected. *The asterisk is not part of the symbol. Identifying valves, letters, numbers, or marks shall replace the asterisk.

COIL, MAGNETIC, BLOWOUT*

CONTACT, ELECTRICAL

For buildups or forms using electrical contacts, see application under CON-NECTOR, RELAY, SWITCH. See DRAFTING PRACTICES.

---

* The broken line — · — indicates where line connection to a symbol is made and is not a part of the symbol.

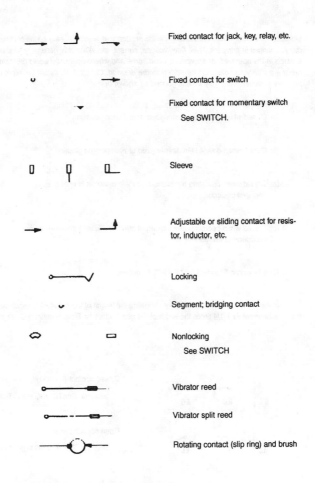

Fixed contact for jack, key, relay, etc.

Fixed contact for switch

Fixed contact for momentary switch
See SWITCH.

Sleeve

Adjustable or sliding contact for resistor, inductor, etc.

Locking

Segment; bridging contact

Nonlocking
See SWITCH

Vibrator reed

Vibrator split reed

Rotating contact (slip ring) and brush

It is standard procedure to show a contact by a symbol which indicates the circuit condition produced when the actuating device is in the nonoperated, or deenergized, position. It may be necessary to add a clarifying note explaining the proper point at which the contact functions – the point where the atuating device (mechanical, electrical, etc.) opens or closes due to changes in pressure, level, flow, voltage, current, etc. When it is necessary to show contacts in the operated, or energized, condition – and where confusion would otherside result – a clarifying note shall be added to the drawing. Contacts for circuit breakers or auxiliary switches, etc., may be designated as shown below:

(a) Closed when device is in energized or operated position

(b) Closed when device is in deenergized or nonoperated position

(aa) Closed when operating mechanism or main device is in energized or operated position

(bb) Closed when operating mechanism of main device is in deenergized or nonoperated position

[See American Standard C37.2-1962 for details.]

In the parallel-line contact, symbols showing the length of the parallel lines shall be approximately 1-1/4 times the width of the gap (except for Time Sequential Closing)

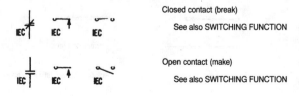

Closed contact (break)

See also SWITCHING FUNCTION

Open contact (make)

See also SWITCHING FUNCTION

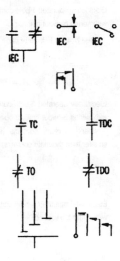

Transfer

    See also SWITCHING FUNCTION

Make-before-break

Application: open contact with time closing (TC) or time delay closing (TDC) feature

Application: closed contact with time opening (TO) or time delay opening (TDO) feature

Time sequential closing

CONTACT, ELECTRICAL

See also RELAY

Contactor symbols are derived from fundamental contact, coil, and mechanical connection symbols, and should be employed to show contactors on complete diagrams. A complete diagram of the actual contactor device is constructed by combining the abovementioned fundamental symbols for mechanical connections, control circuits, etc.

Mechanical interlocking should be indicated by notes.

Manually operated 3-pole contactor

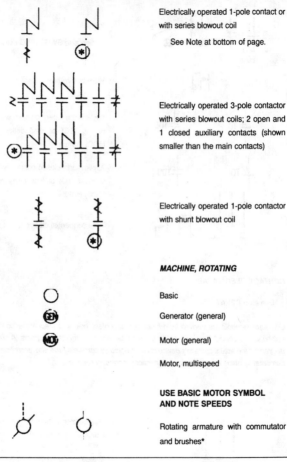

Electrically operated 1-pole contact or with series blowout coil

   See Note at bottom of page.

Electrically operated 3-pole contactor with series blowout coils; 2 open and 1 closed auxiliary contacts (shown smaller than the main contacts)

Electrically operated 1-pole contactor with shunt blowout coil

MACHINE, ROTATING

Basic

Generator (general)

Motor (general)

Motor, multispeed

USE BASIC MOTOR SYMBOL AND NOTE SPEEDS

Rotating armature with commutator and brushes*

Note – The asterisk is not a part of the symbol. Always replace the asterisk by a device designation.

* The broken line — · — indicates where line connection to a symbol is made and is not a part of the symbol.

Field, generator or motor

IEC

Compensating or commutating

IEC

Series

IEC

Shunt, or separately excited

Magnet, permanent

Winding symbols

Motor and generator winding symbols may be shown in the basic circle using the following representation.

1-phase

2-phase

3-phase wye (ungrounded)

3-phase wye (grounded)

3-phase delta

6-phase diametrical

6-phase double delta

Direct-current machines; applications

8-25

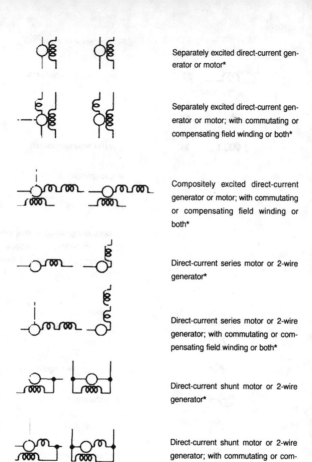

Separately excited direct-current generator or motor*

Separately excited direct-current generator or motor; with commutating or compensating field winding or both*

Compositely excited direct-current generator or motor; with commutating or compensating field winding or both*

Direct-current series motor or 2-wire generator*

Direct-current series motor or 2-wire generator; with commutating or compensating field winding or both*

Direct-current shunt motor or 2-wire generator*

Direct-current shunt motor or 2-wire generator; with commutating or compensating field winding or both*

---

* The broken line — · — indicates where line connection to a symbol is made and is not a part of the symbol.

Direct-current, permanent-magnet field generator or motor*

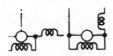 Direct-current, compound motor or 2-wire generator or stabilized shunt motor*

Direct-current compound motor or 2-wire generator or stabilized shunt motor; with commutating or compensating field winding or both*

Direct-current, 3-wire shunt generator*

Direct-current, 3-wire shunt generator; with commutating or compensating field winding or both*

* The broken line — - — indicates where line connection to a symbol is made and is not a part of the symbol.

8-27

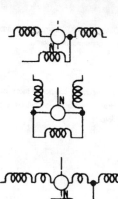

Direct- current, 3-wire compound generator*

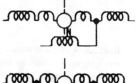

Direct-current, 3-wire compound generator; with commutating or compensating field winding or both*

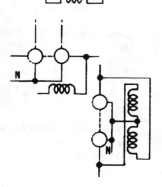

Direct-current balancer, shunt wound*

Alternating-current machines; application

---

* The broken line — - — indicates where line connection to a symbol is made and is not a part of the symbol.

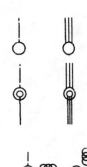

Squirrel-cage induction motor or generator, split-phase induction motor or generator, rotary phase converter or repulsion motor*

Wound-rotor induction motor, synchronous induction motor, induction generator, or induction frequency converter*

Alternating-current series motor*

METER INSTRUMENT

As indicated by the Note, the asterisk is not part of the symbol and should always be replaced with one of the letter combinations listed below, according to the meter's function. This is not necessary if some other identification is provided in the circle and described in the diagram.

A	Ammeter IEC
AH	Ampere-hour
CMA	Contact-making (or breaking) ammeter
CMC	Contact-making (or breaking) clock
CMV	Contact-making (or breaking) voltmeter
CRO	Oscilloscope or cathode-ray oscillograph
DB	DB (decibel) meter
DBM	DMB (decibels referred to 1 milliwatt) meter

---

* The broken line — · — indicates where line connection to a symbol is made and is not a part of the symbol.

DM	Demand meter
DTR	Demand-totalizing relay
F	Frequency meter
G	Glavanometer
GD	Ground detector
I	Indicating
INT	Integrating
μA or	
UA	Microammeter
MA	Milliammeter
NM	Noise meter
OHM	Ohmmeter
OP	Oil pressure
OSCG	Oscillograph string
PH	Phasemeter
PI	Position indicator
PF	Power factor
RD	Recording demand meter
REC	Recording
RF	Reaction factor
SY	Synchroscope
TLM	Telemeter
T	Temperature meter
THC	Thermal converter
TT	Total time
V	Voltmeter
VA	Volt-ammeter

VAR	Varmeter
VARH	Varhour meter
VI	Volume indicator; meter, audio level
VU	Standard volume indicator; meter, audio level
W	Wattmeter
WH	Watthour meter

PATH, TRANSMISSION, CONDUCTOR, CABLE, WIRING

_____ IEC

Guided path, general

The entire group of conductors, or the transmission path required to guide the power or symbol, is shown by a single line. In coaxial and waveguide work, the recognition symbol is employed at the beginning and end of each type of transmission path as well as at intermediate points to clarify a potentially confusing diagram. For waveguide work, the mode may be indicated as well.

_____ IEC

Conductive path or conductor; wire

—//— IEC   ===== IEC

Two conductors or conductive paths of wires

—///— IEC   ≡≡≡ IEC

Three conductors or conductive paths of wires

8-31

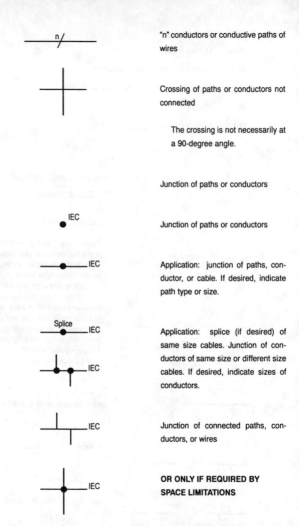

"n" conductors or conductive paths of wires

Crossing of paths or conductors not connected

    The crossing is not necessarily at a 90-degree angle.

Junction of paths or conductors

Junction of paths or conductors

Application: junction of paths, conductor, or cable. If desired, indicate path type or size.

Application: splice (if desired) of same size cables. Junction of conductors of same size or different size cables. If desired, indicate sizes of conductors.

Junction of connected paths, conductors, or wires

OR ONLY IF REQUIRED BY SPACE LIMITATIONS

+	IEC	Positive
–	IEC	Negative

SWITCH

See also FUSE; CONTACT, ELECTRIC; and DRAFTING PRACTICES.

Switch symbols may be constructed of the fundamental symbols for mechanical connections, contacts, etc.

In standard procedure, a switch is represented in the nonoperating or deenergized position. In the case of switches that have two or more positions in which no operating force is applied and for those switches (air-pressure, liquid-level, rate-of-flow, etc.) that may be actuated by a mechanical force, the point at which the switch functions should be described in a clarifying note.

In cases where the basic switch symbols are used in a diagram in the closed position, the terminals must be included for clarity.

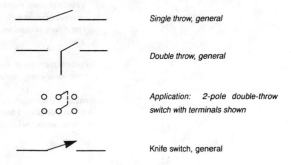

Single throw, general

Double throw, general

Application: 2-pole double-throw switch with terminals shown

Knife switch, general

Pushbutton, momentary or spring return

o̶—̶o   Circuit closing (make)

o̶—l—̶o   Circuit opening (break)

o̶—l—̶o
o   o   Two-circuit

Push-button, maintained or not spring return

 Two-circuit

TRANSFORMER

General

IEC

Either winding symbol may be used. In the following items, the left symbol is used. Additional windings may be shown or indicated by a note. For power transformers use polarity marking H_1, X_1, etc., from American Standard C6.1-1956.

For polarity markings on current and potential transformers, see page 36.

In coaxial and waveguide circuits, this symbol will represent a taper or step transformer without mode change.

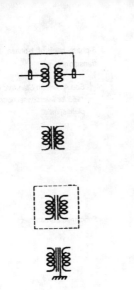

Application: transformer with direct-current connections and mode suppression between two rectangular waveguides

If it is desired especially to distinguish a magnetic-core transformer

Application: shielded transformer with magnetic core shown

Application: transformer with magnetic core shown and with a shield between windings. The shield is shown connected to the frame.

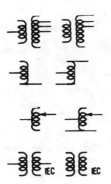

With taps, 1-phase

Autotransformer, 1-phase

Adjustable

1-phase, 2-winding transformer

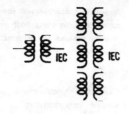

3-phase bank of 1-phase, 2-winding transformer

See American Standard C6.1-1965 for interconnections for complete symbol.

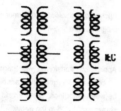

Polyphase transformer

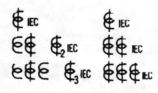

Current transformer(s)

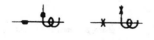

Current transformer with polarity marking. Instantaneous direction of current into one polarity mark responds to current out of the other polarity mark.

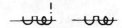 Bushing-type current transformer*

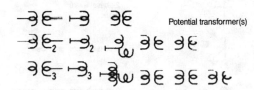 Potential transformer(s)

 Potential transformer with polarity mark. Instantaneous direction of current into one polarity mark corresponds to current out of the other polarity mark.

 Outdoor metering device

Transformer winding connection symbols

For use adjacent to the symbols for the transformer windings.

2-phase 3-wire, grounded

---

* The broken line — · — indicates where line connection to a symbol is made and is not a part of the symbol.

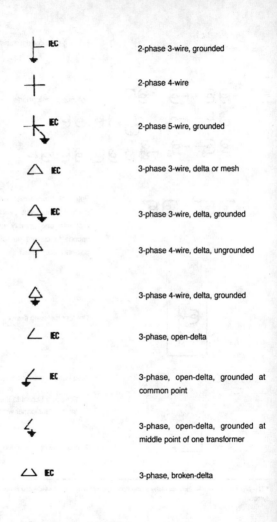

2-phase 3-wire, grounded

2-phase 4-wire

2-phase 5-wire, grounded

3-phase 3-wire, delta or mesh

3-phase 3-wire, delta, grounded

3-phase 4-wire, delta, ungrounded

3-phase 4-wire, delta, grounded

3-phase, open-delta

3-phase, open-delta, grounded at common point

3-phase, open-delta, grounded at middle point of one transformer

3-phase, broken-delta

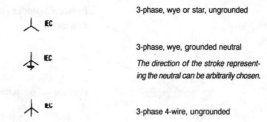

3-phase, wye or star, ungrounded

3-phase, wye, grounded neutral

The direction of the stroke representing the neutral can be arbitrarily chosen.

3-phase 4-wire, ungrounded

MOTOR CONTROLS

Motor controls are not really complicated. Once the fundamental idea is mastered, all types of controls can be figured out and sketched. First, you must become familiar with the common symbols that are used in connection with controls; second, analyze what you want to accomplish with the particular control. If you follow through, your diagram will fit right into place.

There are two common illustrative methods – diagrams and schematics. Both are used, and in practically every case either one will be acceptable to the examiner. However, you should ask the examiner to be certain that you may use either; schematic method, however, presents a clearer picture of what you are trying to accomplish and is easier for the examiner to follow and correct. Whichever method you use take your time and make a clear sketch. There is nothing so irritating to an examiner as trying to determine what your intent was and whether you had any idea what you were doing.

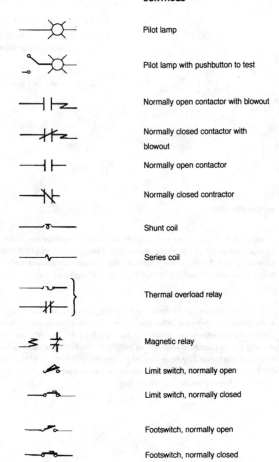

COMMON SYMBOLS FOR MOTOR CONTROLS

Pilot lamp

Pilot lamp with pushbutton to test

Normally open contactor with blowout

Normally closed contactor with blowout

Normally open contactor

Normally closed contractor

Shunt coil

Series coil

Thermal overload relay

Magnetic relay

Limit switch, normally open

Limit switch, normally closed

Footswitch, normally open

Footswitch, normally closed

Vacuum switch, normally open

Vacuum switch, normally closed

Liquid-level switch, normally open

Liquid-level switch, normally closed

Temperature-actuated switch, normally open

Temperature-actuated switch, normally closed

Flow switch, normally open

Flow switch, normally closed

Momentary-contact switch, normally open

Momentary-contact switch, normally closed

Iron-core inductor

Air-core inductor

Single-phase AC motor

3-phase, squirrel-cage motor

2-phase, 4-wire motor

Wound-rotor, 3-phase motor

Armature

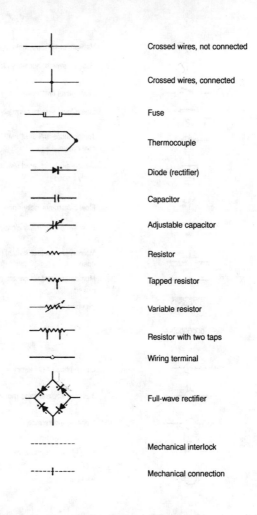

Crossed wires, not connected

Crossed wires, connected

Fuse

Thermocouple

Diode (rectifier)

Capacitor

Adjustable capacitor

Resistor

Tapped resistor

Variable resistor

Resistor with two taps

Wiring terminal

Full-wave rectifier

Mechanical interlock

Mechanical connection

CHAPTER 9
LIGHTING

RECOMMENDED LIGHT LEVELS

Area	Footcandles
Hospital operating table	2500
Factory assembly are	100-200
Accounting office	150
Major league baseball Infield	150
Major league baseball outfield	100
School classroom	60-90
Home kitchen	50-70
Bank lobby	50
Active warehouse storage area	20-30
Inactive warehouse storage area	5
Auditorium	15
Elevator	10
Parking lot	5
Interstate roadway	1.4
Street	.9

LAMP RATINGS

Lamp	Initial Lumen	Mean Lumen
40 W standard incandescent	480	N/A
100 W standard incandescent	1750	N/A
40 W standard fluorescent	3400	3100
100 W tungsten-halogen	1800	1675
100 W mercury-vapor	4000	3000
250 W mercury-vapor	12,000	9800
100 W high-pressure sodium	9500	8500
250 W high-pressure sodium	30,000	27,000
250 W metal-halide	20,000	17,000

LAMP ADVANTAGES AND DISADVANTAGES

Lamp	Advantages	Disadvantages
Incandescent, tungsten-halogen	Low initial cost Simple construction No ballast required Available in many shapes and sizes Requires no warm-up or restart time Inexpensively dimmed Simple maintenance	Low electrical efficiency High operating temperature Short life Bright light source in small space Does not allow large distribution of light
Fluorescent	Available in many shapes and sizes Moderate cost Good electrical efficiency Long life Low shadowing Low operating temperature Short turn-ON delay	Not suited for high-level light in small, highly-concentrated applications Requires ballast Higher initial cost than incandescent lamps Light output and color affected by ambient temperature Expensive to dim
Low-pressure sodium, mercury-vapor, metal-halide, high-pressure	Good electrical efficiency Long life High light output Slightly affected by ambient temperature	May cause color distortion Long start and restart time High initial cost High replacement cost Requires ballast Expensive or not possible to dim Problem starting in cold weather High-socket voltage required

LAMP ELECTRICAL EFFICIENCY

Lamp	Lumen Per Watt*
Incandescent	15-25
Mercury-vapor	50-60
Fluorescent	55-100
Metal-halide	80-125
High-pressure sodium	80-150
Low-pressure sodium	160-200

* exact lumen output depends on size and type of lamp used

COMMON INCANDESCENT LAMPS

Type	Watts	Size (inches)	Volts
A-19	40-100	2-3/8	120-130
F-15	25-60	1-7/8	120
PS-35	300-500	4-3/8	120
PAR-38	115-150	4-3/4	120-130
T-19	40-100	2-3/8	120

COMMON FLUORESCENT LAMPS

Type	Watts	Length (inches)	Size (inches)	Base
T-5	4-13	6-21	5/8	Miniature Bi-pin
T-5	13	21	5/8	Miniature Bi-pin
T-8	15-30	18-36	1	Medium Bi-pin
T-9	20-40	6-16 OD	1-1/8	4-pin
T-10	40	48	1-1/4	Medium Bi-pin
T-12	14-75	15-96	1-1/2	Medium Bi-pin
T-12	60-75	96	1-1/2	Single Bi-pin

BALLAST SOUND RATING

Sound Rating	Noise Level (decibels)	Recommended Application
A	20-24	Broadcasting booths, churches, study areas
B	25-30	Libraries, classrooms, quiet office areas
C	31-36	General office areas, commercial buildings
D	37-42	Retail stores, stock rooms
E	43-48	Light production areas, general outdoor lighting
F	Over 48	Street lighting, heavy production areas

HID LAMP OPERATING CHARACTERISTICS

Lamp	Start Time (minutes)	Restart Time
Low-pressure sodium	6-12	4-12 sec
Mercury-vapor	5-6	3-5 min
Metal-halide	2-5	10-15 min
High-pressure sodium	3-4	30-60 sec

LAMP CHARACTERISTICS SUMMARY

Lamp	Lm/W	Rated Bulb Life (hours)	Color Rendition	Operating Cost
Incandescent	15-25	750-1000	Excellent	Very high
Tungsten-halogen	20-25	1500-2000	Excellent	High
Fluorescent	55-100	7500-24,000	Very good	Average
Low-pressure sodium	190-200	1800	Poor	Low
Mercury-vapor	50-60	16,000-24,000	Depends on type used	Average
Metal-halide	80-125	3000-20,000	Very good	Average
High-pressure sodium	65-115	7500-14,000	Good (golden white)	Low

HID LAMP TROUBLESHOOTING GUIDE

Problem	Possible Cause	Corrective Action
Lamp does not start	Normal end of life	Replace with new lamp operating characteristic
	Loose lamp connection	Re-seat lamp in socket securely. Ensure lamp holder is rigidly mounted and properly spaced
	Defective photocell used for automatic turn ON	Replace defective photocell
	Low line voltage applied applied to circuit	Ensure that voltage is within ±7% of rated voltage
	Cold area or draft hitting lamp	Protect lamp with enclosure. Use a low temperature-rated ballast
	Defective ballast	Replace ballast
Lamp cycles ON and OFF or flickers when ON	Normal end of life operating characteristic	Replace with new lamp
	Poor electrical connection or loose bulb	Check electrical connections and socket contacts
	Line voltage variations	Ensure that voltage is within±7% of rated voltage. Move lamps to separate circuit when lamps are on same circuit as high-power loads. High-over loads cause a voltage dip when turned ON. This voltage dip may cause the lamp to turn OFF
Short lamp life	Wrong wattage lamp or ballast	HID lamps and ballast must be matched in size. Lamp life is shortened when a large-wattage lamp is used. The same size and type ballast must be installed when replacing a ballast
	Power tap set too low for line voltage	Check ballast taps and ensure they are set for the correct supply voltage
	Defective sodium ballast starter	Replace with new lamp
Low light output	Normal end of life operating characteristic	Replace with new lamp
	Dirty lamp or lamp fixture	Keep lamp fixture clean
	Early blackening of bulb caused by incorrect lamp size or ballast	Ensure that lamp size, lamp type, and ballast match

INCANDESCENT LAMP TROUBLESHOOTING GUIDE

Problem	Possible Cause	Corrective Action
Short lamp life	Voltage higher than lamp rating	Ensure that voltage is equal to or less than lamp's rated voltage. Lamp life is decreased when the voltage is higher than the rated voltage. A 5% higher voltage shortens the lamp life by 40%.
	Lamp exposed to rough service conditions or vibration	Replace lamp with one rated for rough service or resistant to vibration
Lamp does not turn ON after new lamp is installed	Fuse blown, poor electrical connection, faulty control switch	Check circuit fuse or CB. Check electrical connections and voltage out of control switch. Replace switch when voltage is present into, but not out of the control switch.

FLUORESCENT LAMP TROUBLESHOOTING GUIDE

Problem	Circuit Type	Possible Cause	Corrective Action
Lamp blinks has shimmering effect during lighting period	All types	Depletion of emission material on electrodes	Replace with new lamp
New lamp blinks	All types	Loose lamp connection	Reseat lamp in socket securely. Ensure lamp holders are rigidly mounted and properly spaced.
		Low voltage applied to circuit	Ensure that voltage is within ±7% of rated voltage
		Cold area or draft hitting lamp	Protect lamp with enclosure. Use a low temperature-rated ballast
	Preheat	Defective or old starter	Replace starter. Lamp life is reduced when starter is not replaced.
Lamp does not light or is slow starting	All types	Lamp failure	Replace lamp. Test faulty lamp in another fixture. Check circuit fuse or CB. Check voltage at fixture.
		Loss of power to the fixture or low voltage	Troubleshoot fluorescent fixture when voltage is present at the fixture. Replace broken or cracked holder. Check for poor wire connection.
	Preheat Rapid-start	Normal end of starter life Failed capacitor in ballast	Replace starter Replace ballast

(continued)

FLUORESCENT LAMP TROUBLESHOOTING GUIDE

Problem	Circuit Type	Possible Cause	Corrective Action
		Lamp not seated in holder	Seat lamp properly in holder. In rapid-start circuits, the holder includes a switch that removes power when the lamp is removed due to high voltage present
Bulb ends remain lighted after switch is turned OFF	Preheat	Starter contacts stuck together	Replace starter. Lamp life is reduced when starter is not replaced
Short lamp life	All types	Frequent turning ON and OFF of lamps	Normal operation is based on one start per three hour period of operation time. Short lamp life must be expected when frequent starting cannot be avoided
		Supply voltage excessive or low	Check the supply voltage against the ballast rating. Short lamp life must be expected when supply voltage is not within ±7% of lamp rating
		Low ambient temperature. Low temperature causes a slow start	Protect lamp with enclosure. Use a low temperature-rated ballast
	Instant-start	One lamp burned out and other burning dimly due to series-start ballast circuit	Replace burned-out lamp. Ballast is damaged when lamp is not replaced
		Wrong lamp type. May be using rapid-start or pre-heat lamp instead of instant-start	Replace lamp with correct type
Light output decreases	All types	Light output decreases over first 100 hours of operation	Rated light output is based on output after 100 hours of lamp operation. Before 100 hours of operation, light output may be as much as 10% higher than normal
		Low circuit voltage	Check the supply voltage against the ballast rating. Short lamp life and low light output must be expected when supply voltage is not within ±7% of lamp rating
		Dirt build-up on lamp and fixture	Clean bulb and fixture

LAMP IDENTIFICATION

Lamp	Letter	Number	Rating
Mercury-vapor	H	33	400 W
		34	1000 W/130 V
		35	700 W
		36	1000 W/265 V
		37	250 W
		38	100 W
		39	175 W
		43	75 W
Self-ballasted mercury-vapor	B	78	750 W/120 V
Metal-halide	M	47	1000W
		48	1500 W
		57	175 W
		58	250 W
		59	400 W
High-pressure sodium	S	50	250 W
		51	400 W
		52	1000 W
		54	100 W
		55	150 W/55 V
		56	150 W/100 V
		62	70 W
		66	200 W
		67	310 W
		68	50 W
		76	35 W

BALLAST LOST		
Lamp	Lamp Rated Wattage (in W)	Ballast Loss (power loss %)
Low-pressure sodium	70	27
	100	25
	150	22
	250	20
	400	15
	1000	7
Mercury-vapor	40	32
	75	25
	100	22
	175	17
	250	16
	400	14
	700	7
	1000	7
Metal-halide	175	20
	250	17
	400	13
	1000	7
	1500	7
High-pressure sodium	50	30
	100	24
	150	22
	250	20
	400	16
	1000	7

RECOMMENDED BALLAST OUTPUT VOLTAGE LIMITS

Ballast	Lamp Size		rms
	Wattage	ANSI Number	Voltage (Volts)
Low-pressure sodium	18	L69	300-325
	35	L70	455-505
	55	L71	455-505
	90	L72	455-525
	135	L73	645-715
	180	L74	645-715
Mercury-vapor	50	H46	225-255
	75	H43	225-255
	100	H38	225-255
	175	H39	225-255
	250	H37	225-255
	400	H33	225-255
	700	H35	405-455
	1000	H36	405-455
Metal-halide	70	M85	210-250
	100	M90	250-300
	150	M81	220-260
	175	M57	285-320
	250	M80	230-270
	250	M58	285-320
	400	M59	285-320
	1000	M47	400-445
	1500	M48	400-445
High-pressure sodium	35	S76	110-130
	50	S68	110-130
	70	S62	110-130
	100	S54	110-130
	150	S55	110-130
	150	S56	200-250
	200	S66	200-230
	250	S50	175-225
	310	S67	155-190
	400	S51	175-225
	1000	S52	420-480

RECOMMENDED SHORT-CIRCUIT CURRENT TEST LIMITS

Ballast	Lamp Size		Short-Circuit
	Wattage	ANSI Number	Current (Amps)
Low-pressure sodium	18	L69	.30-.40
	35	L70	.52-.78
	55	L71	.52-.78
	90	L72	.8-1.2
	135	L73	.8-1.2
	180	L74	.8-1.2
Mercury-vapor	50	H46	.85-1.15
	75	H43	.95-1.70
	100	H38	1.10-2.00
	175	H39	2.0-3.6
	250	H37	3.0-3.8
	400	H33	4.4-7.9
	700	H35	3.9-5.85
	1000	H36	5.7-9.0
Metal-halide	70	M85	.85-1.30
	100	M90	1.15-1.76
	150	M81	1.75-2.60
	175	M57	1.5-1.90
	250	M80	2.9-4.3
	250	M58	2.2-2.85
	400	M59	3.5-4.5
	1000	M47	4.8-6.15
	1500	M48	7.4-9.6
High-pressure sodium	35	S76	.85-1.45
	50	S68	1.5-2.3
	70	S62	1.6-2.9
	100	S54	2.45-3.8
	150	S55	3.5-5.4
	150	S56	2.0-3.0
	200	S66	2.50-3.7
	250	S50	3.0-5.3
	310	S67	3.8-5.7
	400	S51	5.0-7.6
	1000	S52	5.5-8.1

LIGHT SOURCE CHARACTERISTICS

	Incandescent, Including Tungsten	Fluorescent	High-Intensity Discharge			
			Mercury Vapor (Self-Ballasted)	Metal Halide	High-Pressure Sodium (Improved Color)	Low-Pressure Sodium
Wattages (lamp only)	15-1500	15-219	40-1000	175-1000	70-1000	35-180
Life[a] (hr)	750-12,000	7500-24,000	16,000-15,000	1500-15,000	24,000 (10,000)	18,000
Efficacy[a] (lumens/W) lamp only	15-25	55-100	50-60 (20-25)	80-100	75-140 (67-112)	Up to 180
Lumen maintenance	Fair to excellent	Fair to excellent	Very good (good)	Good	Excellent	Excellent
Color rendition	Excellent	Good to excellent	Poor to excellent	Very good	Fair	Good
Light direction control	Very good to excellent	Fair	Very good	Very good	Very good	Fair
Source size	Compact	Extended	Compact	Compact	Compact	Extended
Relight time	Immediate	Immediate	3-10 min	10-20 min	Less than 1 min	Immediate
Comparative fixture cost	Low: simple fixtures	Moderate	Higher than incandescent and fluorescent	Generally higher than mercury	High	High
Comparative operating cost	High: short life and low efficiency	Lower than incandescent	Lower than incandescent	Lower than mercury	Lowest of HID types	Low
Auxiliary equipment needed	Not needed	Needed: medium cost	Needed high cost	Needed: high cost	Needed: high cost	Needed: high cost

[a]Life and efficacy ratings subject to revision. Check manufacturers' data for latest information.

NOTES

CHAPTER 10
CONVERSION FACTORS &
UNITS OF MEASUREMENT

COMMONLY USED CONVERSION FACTORS

Multiply	By	To Obtain
Acres	43,560	Square feet
Acres	1.562×10^{-3}	Square miles
Aore-Feet	43,560	Cubic feet
Amperes per sq cm	6.452	Amperes per sq in.
Amperes per sq in	0.1550	Amperes per sq cm
Ampere-Turns	1.257	Gilberts
Ampere-Turns per cm	2.540	Ampere-turns per in.
Ampere-Turns per in	0.3937	Ampere-turns per cm
Atmospheres	76.0	Cm of mercury
Atmospheres	29.92	Inches of mercury
Atmospheres	33.90	Feet of water
Atmospheres	14.70	Pounds per sq in.
British termal units	252.0	Calories
British thermal units	778.2	Foot-pounds
British thermal units	3.960×10^{-4}	Horsepower-hours
British termal units	0.2520	Kilogram-calories
British thermal units	107.6	Kilogram-meters
British termal units	2.931×10^{-4}	Kilowatt-hours
British thermal units	1,055	Watt-seconds
B.t.u. per hour	2.931×10^{-4}	Kilowatts
B.t.u. per minute	2.359×10^{-2}	Horsepower
B.t.u. per minute	1.759×10^{-2}	Kilowatts
Bushels	1.244	Cubic feet
Centimeters	0.3937	Inches
Circular mils	5.067×10^{-6}	Square centimeters
Circular mils	0.7854×10^{-6}	Square inches

Multiply	By	To Obtain
Circular mils	0.7854	Square mils
Cords	128	Cubic feet
Cubic centimeters	6.102×10^{-6}	Cubic inches
Cubic feet	0.02832	Cubic meters
Cubic feet	7.481	Gallons
Cubic feet	28.32	Liters
Cubic inches	16.39	Cubic centimeters
Cubic meters	35.31	Cubic feet
Cubic meters	1.308	Cubic yards
Cubic yards	0.7646	Cubic meters
Degrees (angle)	0.01745	Radians
Dynes	2.248×10^{-6}	Pounds
Ergs	1	Dyne-centimeters
Ergs	7.37×10^{-6}	Foot-pounds
Ergs	10^{-7}	Joules
Farads	10^{6}	Microfarads
Fathoms	6	Feet
Feet	30.48	Centimeters
Feet of water	.08826	Inches of mercury
Feet of water	304.8	Kg per square meter
Feet of water	62.43	Pounds per square ft
Feet of water	0.4335	Pounds per square ir
Foot-pounds	1.285×10^{-2}	British thermal units
Foot-pounds	5.050×10^{-7}	Horsepower-hours
Foot-pounds	1.356	Joules
Foot-pounds	0.1383	Kilogram-meters
Foot-pounds	3.766×10^{-7}	Kilowatt-hours
Gallons	0.1337	Cubic feet
Gallons	231	Cubic inches
Gallons	3.785×10^{-3}	Cubic meters
Gallons	3.785	Liters
Gallons per minute	2.228×10^{-3}	Cubic feet per sec.
Gausses	6.452	Lines per square in.
Gilberts	0.7958	Ampere-turns
Henries	10^{3}	Millihenries
Horsepower	42.41	Btu per min
Horsepower	2,544	Btu per hour
Horsepower	550	Foot-pounds per sec

COMMONLY USED CONVERSION FACTORS (CONT.)

Multiply	By	To Obtain
Horsepower	33,000	Foot-pounds per min
Horsepower	1.014	Horsepower (metric)
Horsepower	10.70	Kg calories per min
Horsepower	0.7457	Kilowatts
Horsepower (boiler)	33,520	Btu per hour
Horsepower-hours	2,544	British thermal units
Horsepower-hours	1.98×10^6	Foot-pounds
Horsepower-hours	2.737×10^5	Kilogram-meters
Horsepower-hours	0.7457	Kilowatt-hours
Inches	2.540	Centimeters
Inches of mercury	1.133	Feet of water
Inches of mercury	70.73	Pounds per square ft.
Inches of mercury	0.4912	Pounds per square in.
Inches of water	25.40	Kg per square meter
Inches of water	0.5781	Ounces per square in.
Inches of water	5.204	Pounds per square ft
Joules	9.478×10^{-4}	British thermal units
Joules	0.2388	Calories
Joules	10^7	Ergs
Joules	0.7376	Foot-pounds
Joules	2.778×10^{-7}	Kilowatt-hours
Joules	0.1020	Kilogram-meters
Joules	1	Watt-seconds
Kilograms	2.205	Pounds
Kilogram-calories	3.968	British thermal units
Kilogram meters	7.233	Foot-pounds
Kg per square meter	3.281×10^{-3}	Feet of water
Kg per square meter	0.2048	Pounds per square ft
Kg per square meter	1.422×10^{-3}	Pounds per square in.
Kilolines	10^3	Maxwells
Kilometers	3.281	Feet
Kilometers	0.6214	Miles
Kilowatts	56.87	Btu per min
Kilowatts	737.6	Foot-pounds per sec
Kilowatts	1.341	Horsepower
Kilowatts-hours	3.415	British thermal units
Kilowatts-hours	2.655×10^6	Foot-pounds
Knots	1.152	Miles

Multiply	By	To Obtain
Liters	0.03531	Cubic feet
Liters	61.02	Cubic inches
Liters	0.2642	Gallons
Log N_e or in N	0.4343	Log_{10} N
Log N	2.303	Log_e N or in N
Lumens per square ft	1	Footcandles
Maxwells	10^{-3}	Kilolines
Megalines	10^6	Maxwells
Megaohms	10^6	Ohms
Meters	3.281	Feet
Meters	39.37	Inches
Meter-kilograms	7.233	Pound-feet
Microfarads	10^{-6}	Farads
Microhms	10^{-6}	Ohms
Microhms per cm cube	0.3937	Microhms per in. cube
Microhms per cm cube	6.015	Ohms per mil foot
Miles	5,280	Feet
Miles	1.609	Kilometers
Miner's inches	1.5	Cubic feet per min
Ohms	10^{-6}	Megohms
Ohms	10^6	Microhms
Ohms per mil foot	0.1662	Microhms per cm cube
Ohms per mil foot	0.06524	Microhms per in. cube
Poundals	0.03108	Pounds
Pounds	32.17	Poundals
Pound-feet	0.1383	Meter-Kilograms
Pounds of water	0.01602	Cubic feet
Pounds of water	0.1198	Gallons
Pounds per cubic foot	16.02	Kg per cubic meter
Pounds per cubic foot	5.787×10^{-4}	Pounds per cubic in.
Pounds per cubic inch	27.68	Grams per cubic cm
Pounds per cubic inch	2.768×10^{-4}	Kg per cubic meter
Pounds per cubic inch	1.728	Pounds per cubic ft
Pounds per square foot	0.01602	Feet of water
Pounds per square foot	4.882	Kg per square meter
Pounds per square foot	6.944×10^{-3}	Pounds per sq. in.
Pounds per square inch	2.307	Feet of water
Pounds per square inch	2.036	Inches of mercury

COMMONLY USED CONVERSION FACTORS (CONT.)

Multiply	By	To Obtain
Pounds per square inch	703.1	Kg per square meter
Radians	57.30	Degrees
Square centimeters	1.973×10^5	Circular mils
Square Feet	2.296×10^{-5}	Acres
Square Feet	0.09290	Square meters
Square inches	1.273×10^6	Circular mils
Square inches	6.452	Square centimeters
Square Kilometers	0.3861	Square miles
Square meters	10.76	Square feet
Square miles	640	Acres
Square miles	2.590	Square kilometers
Square Milimeters	1.973×10^3	Circular mils
Square mils	1.273	Circular mils
Tons (long)	2,240	Pounds
Tons (metric)	2,205	Pounds
Tons (short)	2,000	Pounds
Watts	0.05686	Btu per minute
Watts	10^7	Ergs per sec
Watts	44.26	Foot-pounds per min.
Watts	1.341×10^{-3}	Horsepower
Watts	14.34	Calories per min.
Watts-hours	3.412	British thermal units
Watts-hours	2,655	Footpounds
Watts-hours	1.341×10^{-3}	Horsepower-hours
Watts-hours	0.8605	Kilogram-calories
Watts-hours	376.1	Kilogram-meters
Webers	10^8	Maxwells

ELECTRICAL PREFIXES

Prefixes
Prefixes are used to avoid long expressions of units that are smaller and larger than the base unit. See Common Prefixes. For example, sentences 1 and 2 do not use prefixes. Sentences 3 and 4 use prefixes.
1. A solid-state device draws 0.000001 amperes (A).
2. A generator produces 100,000 watts (W).
3. A solid-state device draws 1 microampere (μA).
4. A generator produces 100 kilowatts (kW).

Converting Units
To convert between different units, the decimal point is moved to the left or right, depending on the unit. See Conversion Table. For example, an electronic circuit has a current flow of .000001 A. The current value is converted to simplest terms by moving the decimal point six places to the right to obtain 1.0 μA (from Conversion Table).

$$.000001. \text{ A} = 1.0 \ \mu\text{A}$$

Move decimal point 6 places to right

Common Electrical Quantities
Abbreviations are used to simplify the expression of common electrical quantities. See Common Electrical Quantities. For example, milliwatt is abbreviated mW, kilovolt is abbreviated kV, and ampere is abbreviated A.

COMMON PREFIXES

Symbol	Prefix	Equivalent
G	giga	1,000,000,000
M	mega	1,000,000
k	kilo	1000
base unit	—	1
m	milli	.001
m	micro	.000001
n	nano	.000000001

COMMON ELECTRICAL QUANTITIES

Variable	Name	Unit of Measure and Abbreviation
E	voltage	volt - V
I	current	ampere - A
R	resistance	ohm - Ω
P	power	watt - W
P	power (apparent)	volt-ampere - VA
C	capacitance	farad - F
L	inductance	henry - H
Z	impedance	ohm - Ω
G	conductance	siemens - S
f	frequency	hertz - Hz
T	period	second - s

CONVERSION TABLE

Initial Units	Final Units						
	giga	mega	kilo	base unit	milli	micro	nano
giga	—	3R	6R	9R	12R	15R	18R
mega	3L	—	3R	6R	9R	12R	15R
kilo	6L	3L	—	3R	6R	9R	12R
base unit	9L	6L	3L	—	3R	6R	9R
milli	12L	9L	6L	3L	—	3R	6R
micro	15L	12L	9L	6L	3L	—	3R
nano	18L	15L	12L	9L	6L	3L	—

10-6

ELECTRICAL ABBREVIATIONS

Abbrev.	Term	Abbrev.	Term
A	Amps; armature; anode; ammeter	K	Kilo; cathode
Ag	Silver	L	Line; load
ALM	Alarm	LB-FT	Pounds per feet
AM	Ammeter	LB-IN	Pounds per inch
ARM	Armature	LRC	Locked rotor current
Au	Gold	M	Motor; motor starter contacts
BK	Black	MED	Medium
BL	Blue	N	Nirth
BR	Brown	NC	Normally closed
C	Celsius; centigrade	NO	Normally opened
CAP	Capacitor	NTDF	Nontime-delay fuse
CB	Circuit breaker	O	Orange
CCW	Counterclockwise	OCPD	Overcurrent protection device
CONT	Continuous	OL	Overloads
CPS	Cycles per second	OZ/IN	Ounces per inch
CR	Control relay	P	Power consumed
CT	Current transformer	PSI	Pounds per square inch
CW	Clockwise	PUT	Pull-up torque
D	Diameter	R	Resistance; radius; red; reverse
DP	Double-pole	REV	Reverse
DPDT	Double-pole, double-throw	RPM	Revolutions per minute
EMF	Electromotive force	S	Switch; series; slow; south
F	Fahrenheit; forward; fast	SCR	Silicon controlled rectifier
F	Field; forward	SF	Service factor
FLC	Full-load current	SP	Single-pole
FLT	Full-load torque	SPDT	Single-pole; double-throw
FREQ	Frequency	SPST	Single-pole; single-throw
FS	Float switch	SW	Switch
FTS	Foot switch	T	Terminal; torque
FWD	Forward	TD	Time delay
G	Green; gate	TDF	Time-delay fuse
GEN	Generator	TEMP	Temperature
GY	Gray	V	Volts; violet
H	Transformer, primary side	VA	Voltamps
HP	Horsepower	VAC	Volts alternating current
I	Current	VDC	Volts direct current
IC	Intergrated circuit	W	White; watt
INT	Intermediate; interrupt	W/	With
ITB	Inverse time breaker	X	Transformer secondary side
ITCB	Instantaneous trip circuit breaker	Y	Yellow

TEMPERATURE CONVERSIONS

°C	°F	°C	°F	°C	°F
10000	18032	430	806	200	392.0
9500	17132	420	788	195	383.0
9000	16232	410	770	190	374.0
8500	15332	400	752	185	365.0
8000	14432	395	743	180	356.0
7500	13532	390	734	175	347.0
7000	12632	385	725	170	338.0
6500	11732	380	716	165	329.0
6000	10832	375	707	160	320.0
5500	9932	370	698	155	311.0
5000	9032	365	689	150	302.0
4500	8132	360	680	145	293.0
4000	7232	355	671	140	284.0
3500	6332	350	662	135	275.0
3000	5432	345	653	130	266.0
2500	4532	340	644	125	257.0
2000	3632	335	635	120	248.0
1500	2732	330	626	115	239.0
1000	1832	325	617	110	230.0
950	1742	320	608	105	221.0
900	1652	315	599	100	212.0
850	1562	310	590	99	210.2
800	1472	305	581	98	208.4
750	1382	300	572	97	206.6
700	1292	295	563	96	204.8
650	1202	290	554	95	203.0
600	1112	285	545	94	201.2
590	1094	280	536	93	199.4
580	1076	275	527	92	197.6
570	1058	270	518	91	195.8
560	1040	265	509	90	194.0
550	1022	260	500	89	192.2
540	1004	255	491	88	190.4
530	986	250	482	87	188.6
520	968	245	473	86	186.8
510	950	240	464	85	185.0
500	932	235	455	84	183.2
490	914	230	446	83	181.4
480	896	225	437	82	179.6
470	878	220	428	81	177.8
460	860	215	419	80	176.0
450	842	210	410	79	174.2
440	824	205	401	78	172.4

°C = Degrees Celsius. 1 unit is 1/100 of the difference between the temperature of melting ice and boiling water at standard temperature and pressure.

°F = Degrees Fahrenheit. 1 unit is 1/180 of the difference between the temperature of melting ice and boiling water at standard temperature and pressure.

TEMPERATURE CONVERSIONS (CONT.)

°C	°F	°C	°F	°C	°F
77	170.6	34	93.2	-9	15.8
76	168.8	33	91.4	-10	14.0
75	167.0	32	89.6	-11	12.2
74	165.2	31	87.8	-12	10.4
73	163.4	30	86.0	-13	8.6
72	161.6	29	84.2	-14	6.8
71	159.8	28	82.4	-15	5.0
70	158.0	27	80.6	-16	3.2
69	156.2	26	78.8	-17	1.4
68	154.4	25	77.0	-18	-0.4
67	152.6	24	75.2	-19	-2.2
66	150.8	23	73.4	-20	-4.0
65	149.0	22	71.6	-21	-5.8
64	147.2	21	69.8	-22	-7.6
63	145.4	20	68.0	-23	-9.4
62	143.6	19	66.2	-24	-11.2
61	141.8	18	64.4	-25	-13.0
60	140.0	17	62.6	-26	-14.8
59	138.2	16	60.8	-27	-16.6
58	136.4	15	59.0	-28	-18.4
57	134.6	14	57.2	-29	-20.2
56	132.8	13	55.4	-30	-22.0
55	131.0	12	53.6	-31	-23.8
54	129.2	11	51.8	-32	-25.6
53	127.4	10	50.0	-33	-27.4
52	125.6	9	48.2	-34	-29.2
51	123.8	8	46.4	-35	-31.0
50	122.0	7	44.6	-36	-32.8
49	120.2	6	42.8	-37	-34.6
48	118.4	5	41.0	-38	-36.4
47	116.6	4	39.2	-39	-38.2
46	114.8	3	37.4	-40	-40.0
45	113.0	2	35.6	-50	-58.0
44	111.2	1	33.8	-60	-76.0
43	109.4	0	32.0	-70	-94.0
42	107.6	-1	30.2	-80	-112.0
41	105.8	-2	28.4	-90	-130.0
40	104.0	-3	26.6	-100	-148.0
39	102.2	-4	24.8	-125	-193.0
38	100.4	-5	23.0	-150	-238.0
37	98.6	-6	21.2	-200	-328.0
36	96.8	-7	19.4	-250	-418.0
35	95.0	-8	17.6	-273	-459.4

$$°C = 5/6 \ (°F - 32) \qquad °F = 9/5 \ °C + 32$$
$$\text{Absolute Zero} = 0K = -273.16°C = -459.69°F$$

K = Kelvin (Absolute temperature). This scale is based on the average kinetic energy per molecule of a perfect gas and uses the same size unit as the Celsius scale, but the degree symbol (°) is not used. Zero (0K) on the scale is the temperature at which a perfect gas has lost all of its energy.

SOUND INTENSITIES

Decibels	Degree	Loudness or Feeling
225	Deafening	12" cannon @ 12ft, in front & below
194		Saturn rocket, 50# of TNT @ 10'
140		Artillery fire, jet aircraft, ram jet
130		Threshold of pain
		>130 causes immediate ear damage
		Propeller aircraft at 5 meters
		Hydraulic press, pneumatic rock drill
120		Thunder, diesel engine room
		Nearby riviter
110		Close to a train, ball mill
100	Very Loud	Boiler factory, home lawn mower
		Car horn at 5 meters, wood saw
90		Symphony or a band
		>90 regularly can cause ear damage
		Noisy factory
		Truck without muffler
80	Loud	Inside a high speed auto
		Police whistle, electric shaver
		Noisy office, alarm clock
70		Average radio
		Normal street noise
60	Moderate	Normal conversation, close up
50		Normal office noise, quiet stream
45		To awaken a sleeping person
40	Faint	Normal private office noise
		Residential neighborhood, no cars
30		Quiet conversation, recording studio
20	Very Faint	Inside an empty theater
		Ticking of a watch
		Rustle of leaves
		Whisper
10		Sound proof room
		Threshold of hearing
0		Absolute silence

Sound intensities are typically measured in decibels (db). A decibel is defined as 10 times the logarithm of the power ratio (power ratio is the ratio of the intensity of the sound to the intensity of an arbitrary standard point). Normally a change of 1 db is the smallest volume change detectable by the human ear.

Sound intensity is also defined in terms of energy (erg) transmitted per second over a 1 square centimeter surface. This energy is proportional to the velocity of propagation of the sound. The energy density is $erg/cm^3 = 2\pi^2$ x density in g/cm^3 x frequency^2 in hz x amplitude^2 in cm.

SOUND INTENSITIES (CONT.)

Permissible Noise Exposures

Hours Duration per Day	Sound Level in Decibels (Slow Response)
8	90
6	92
4	95
3	97
2	100
1.5	102
1	105
0.5	110
0.25	115

The above restrictions are based on the *Occupational Safety and Health Act of 1970*. That Code basically states that if the above exposures are exceeded, then hearing protection must be worn. Note that these are based on the "A scale" of standard sound level meter at slow response and will change is some other standard is used. See the *OSHA Section 1910.95* for additional details on the differences.

Perception of Changes in Sound

Sound Level Change in Decibels	Perception
3	Barely perceptible
5	Clearly perceptible
10	Twice as loud

Note that the sound level scale in decibels is a logarithmic rather than linear scale. A sound level change of 3 decibels is double (or half) of the previous power level. The ear registers this as just noticeable. A change in power level of 10 decibels is a power change of 10 times, and the ear judges this as only twice (or half) as loud.

These relationships do not hold true at all power levels or at all frequencies, as the ear is a very non-linear device. See Fletcher-Munson hearing curves published in books on hearing.

Some human ears can hear sounds in the frequency range of 20Hz to 20,000Hz, however, the hearing for most people is limited to about 30Hz to 15,000Hz.

RIGHT TRIANGLE TRIG FORMULAS

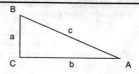

A, B, C = Angles   a, b, c = Distances

$$\sin A = \frac{a}{c}, \quad \cos A = \frac{b}{c}, \quad \tan A = \frac{a}{b}$$

$$\cot A = \frac{b}{a}, \quad \sec A = \frac{c}{b}, \quad \operatorname{cosec} A = \frac{c}{a}$$

<u>Given a and b, Find A, B, and c</u>

$$\tan A = \frac{a}{b} = \cot B, \quad c = \sqrt{a^2 + b^2} = a\sqrt{1 + \frac{b^2}{a^2}}$$

<u>Given a and c, Find A, B, and b</u>

$$\sin A = \frac{a}{c} = \cos B, \quad b = \sqrt{(c + a)(c - a)} = c\sqrt{1 - \frac{a^2}{c^2}}$$

<u>Given A and a, Find B, b, c</u>

$$B = 90° - A, \quad b = a \cot A, \quad c = \frac{a}{\sin A}$$

<u>Given A and b, Find B, a, c</u>

$$B = 90° - A, \quad a = b \tan A, \quad c = \frac{b}{\cos A}$$

<u>Given A and c, Find B, a, b</u>

$$B = 90° - A, \quad a = c \sin A, \quad b = c \cos A$$

OBLIQUE TRIANGLE FORMULAS

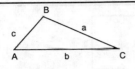 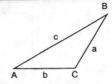

Given A, B and a. Find b, C, and c

$$b = \frac{a \sin B}{\sin A} \quad , \quad C = 180° - (A + B) \quad , \quad c = \frac{a \sin C}{\sin A}$$

Given A, a and b. Find B, C, and c

$$\sin B = \frac{b \sin A}{a} \quad , \quad C = 180° - (A + B) \quad , \quad c = \frac{a \sin C}{\sin A}$$

Given a, b and C. Find A, B, and c

$$A + B = 180° - C \quad , \quad c = \frac{a \sin C}{\sin A}$$

$$\tan A = \frac{a \sin C}{b - (a \cos C)}$$

Given a, b and c. Find A, B, and C

$$s = \frac{a + b + c}{2} \quad , \quad \sin \tfrac{1}{2} A = \sqrt{\frac{(s - b)(s - c)}{bc}}$$

$$\sin \tfrac{1}{2} B = \sqrt{\frac{(s - a)(s - c)}{ac}} \quad , \quad C = 180° - (A + B)$$

Given a, b and c. Find Area

$$s = \frac{a + b + c}{2} \quad , \quad Area = \sqrt{S(s - a)(s - b)(s - c)}$$

$$Area = \frac{bc \sin A}{2} \quad , \quad Area = \frac{a^2 \sin B \sin C}{2 \sin A}$$

PLANE FIGURE FORMULAS

Rectangle

If square, a = b

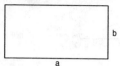

Area = ab

Perimeter = 2(a + b) , Diagonal = $\sqrt{a^2 + b^2}$

Parallelogram

All sides are parallel
θ = degrees

Area = ah = ab sin θ , Perimeter = 2 (a + b)

Trapezoid

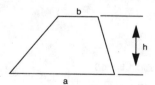

Area = $\frac{(a + b)}{2}$ h

Perimeter = Sum of lengths of sides

Quadrilateral

θ = degrees

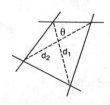

Area = $\frac{d_1 \times d_2 \times \sin \theta}{2}$

PLANE FIGURE FORMULAS (CONT.)

Trapezium

a to g = lengths

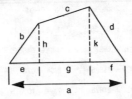

$Perimeter = a + b + c + d$

$$Area = \frac{(h+k)\,g + e\,h + f\,k}{2}$$

Equilateral Triangle

a = all sides equal

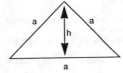

$Perimeter = 3a$, $h = \frac{a\sqrt{3}}{2} = 0.866\,a$

$$Area = a^2\frac{\sqrt{3}}{4} = 0.433\,a^2$$

Annulus

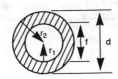

$Area = 0.7854\,(d^2 - f^2)$

$Area = \pi\,(r_1 + r_2)\,(r_2 - r_1)$

DECIMAL EQUIVALENTS OF 8THS, 16THS, 32NDS, 64THS

8ths	32nds	64ths	64ths
1/8 = .125	1/32 = .03125	1/64 = 0.15625	33/64 = .515625
1/4 = .250	3/32 = .09375	3/64 = .046875	35/64 = .546875
3/8 = .375	5/32 = .15625	5/64 = .078125	37/64 = .57812
1/2 = .500	7/32 = .21875	7/64 = .109375	39/64 = .609375
5/8 = .625	9/32 = .28125	9/64 = .140625	41/64 = .640625
3/4 = .750	11/32 = .34375	11/64 = .171875	43/64 = .671875
7/8 = .875	13/32 = .40625	13/64 = .203128	45/64 = .703125
16ths	15/32 = .46875	15/64 = .234375	47/64 = .734375
1/16 = .0625	17/32 = .53125	17/64 = .265625	49/64 = .765625
3/16 = .1875	19/32 = .59375	19/64 = .296875	51/64 = 3796875
5/16 = .3125	21/32 = .65625	21/64 = .328125	53/64 = .828125
7/16 = .4375	23/32 = .71875	23/64 = .359375	55/64 = .859375
9/16 = .5625	25/32 = .78125	25/64 = .390625	57/64 = .890625
11/16 = .6875	27/32 = .84375	27/64 = .421875	59/64 = .921875
13/16 = .8125	29/32 = .90625	29/64 = .453125	61/64 = .953125
15/16 = .9375	31/32 = .96875	31/64 = .484375	63/64 = .984375

COMMON ENGINEERING UNITS AND THEIR RELATIONSHIP

Quantity	SI Metric Units/Symbols	Customary Units	Relationship of Units
Acceleration	meters per second squared (m/s²)	feet per second squared (ft/s²)	$m/s^2 = ft/s^2 \times 3.281$
Area	square meter (m²) square millimeter (mm²)	square foot (ft²) square inch (in²)	$m^2 = ft^2 \times 10.764$ $mm^2 = in^2 \times 0.00155$
Density	kilograms per cubic meter (kg/m³) grams per cubic centimeter (g/cm³)	pounds per cubic foot (lb/ft³) pounds per cubic inch (lb/in³)	$kg/m^3 = lb/ft^2 \times 16.02$ $g/cm^3 = lb/in^3 \times 0.036$
Work	Joule (J)	foot pound force (ft lbf or ft lb)	$J = ft\ lbf \times 1.356$
Heat	Joule (J)	British thermal unit (Btu) Calorie (Cal)	$J = Btu \times 1.055$ $J = cal \times 4.187$
Energy	kilowatt (kW)	Horsepower (HP)	$kW = HP \times 0.7457$
Force	Newton (N) Newton (N)	Pound-force (lbf, lb · f, or lb) kilogram-force (kgf, kg · f, or kp)	$N = lbf \times 4.448$ $N = \dfrac{kgf}{9.807}$
Length	meter (m) millimeter (mm)	foot (ft) inch (in)	$m = ft \times 3.281$ $mm = \dfrac{in}{25.4}$
Mass	kilogram (kg) gram (g)	pound (lb) ounce (oz)	$kg = lb \times 2.2$ $g = \dfrac{oz}{28.35}$
Stress	Pascal = Newton per second (Pa = N/s)	pounds per square inch (lb/in² or psi)	$Pa = lb/in^2 \times 6{,}895$
Temperature	degree Celsius (°C)	degree Fahrenheit (°F)	$°C = \dfrac{°F - 32}{1.8}$
Torque	Newton meter (N · m)	foot-pound (ft lb) inch-pound (in lb)	$N \cdot m = ft\ lbf \times 1.356$ $N \cdot m = in\ lbf \times 0.113$
Volume	cubic meter (m³) cubic centimeter (cm³)	cubic foot (ft³) cubic inch (in³)	$m^3 = ft^3 \times 35.314$ $cm^3 = \dfrac{in^3}{16.387}$

COMMONLY USED GEOMETRICAL RELATIONSHIPS

Diameter of a circle x 3.1416 = Circumference.

Radius of a circle x 6.283185 = Circumference.

Square of the radius of a circle x 3.1416 = Area.

Square of the diameter of a circle x 0.7854 = Area.

Square of the circumference of a circle x 0.07958 = Area.

Half the circumference of a circle x half its diameter = Area.

Circumference of a circle x 0.159155 = Radius.

Square root of the area of a circle x 0.56419 = Radius.

Circumference of a circle x 0.31831 = Diameter.

Square root of the area of a circle x 1.12838 = Diameter.

Diameter of a circle x 0.866 = Side of an inscribed equilateral triangle.

Diameter of a circle x 0.7071 = Side of an inscribed square.

Circumference of a circle x 0.225 = Side of an inscribed square.

Circumference of a circle x 0.282 = Side of an equal square.

Diameter of a circle x 0.8862 = Side of an equal square.

Base of a triangle x one-half the altitude = Area.

Multiplying both diameters and .7854 together = Area of an ellipse.

Surface of a sphere x one-sixth of its diameter = Volume.

Circumference of a sphere x its diameter = Surface.

Square of the diameter of a sphere x 3.1416 = Surface.

Square of the circumference of a sphere x 0.3183 = Surface.

Cube of the diameter of a sphere x 0.5236 = Volume.

Cube of the circumference of a sphere x 0.016887 = Volume.

Radius of a sphere x 1.1547 = Side of an inscribed cube.

Diameter of a sphere divided by $\sqrt{3}$ = Side of an inscribed cube.

Area of its base x one-third of its altitude = Volume of a cone or
pyramid whether round, square or triangular.

Area of one of its sides x 6 = Surface of the cube.

Altitude of trapezoid x one-half the sum of its parallel sides = Area.

CHAPTER 11
MATERIALS & TOOLS

FEET OF CABLE ON A REEL

The following formula can be used to accurately determine the number of feet of rope or cable that is smoothly wound on a drum or reel (A, B, and C are in inches):

Length in Feet = A x [A + B] x C x K

Values of K for above equation

Rope Diameter Inches	Value of K	Rope Diameter Inches	Value of K
1/4	3.29	1-1/8	0.191
5/16	2.21	1-1/4	0.152
3/8	1.58	1-3/8	0.127
7/16	1.19	1-1/2	0.107
1/2	0.925	1-5/8	0.0886
9/16	0.741	1-3/4	0.0770
5/8	0.607	1-7/8	0.0675
3/4	0.428	2	0.0597
7/8	0.308	2-1/8	0.0532
1	0.239	2-1/4	0.0476

PULL ANGLE vs STRENGTH LOSS

The load carrying capacity of a cable, rope, sling, etc decreases by the factor K as the and θ increases.

θ	K	θ	K
5	0.9962	45	0.7071
10	0.9848	50	0.6428
15	0.9659	55	0.5736
20	0.9397	60	0.5000
25	0.9063	65	0.4226
30	0.8660	70	0.3420
35	0.8792	75	0.2588
40	0.7660		

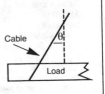

BOLT TORQUE SPECIFICATIONS

Bolt Size Inches	Coarse Thread /Inch	SAE 0-1-2 74,000 psi Low Carbon Steel	SAE Grad 3 100,000 psi Med. Carbon Steel	SAE Grade 5 120,000 psi Med. Carbon Steel
		Standard Dry Torque in Foot-Pounds		
1/4	20	6	9	10
5/16	18	12	17	19
3/8	16	20	30	33
7/16	14	32	47	54
1/2	13	47	69	78
9/16	12	69	103	114
5/8	11	96	145	154
3/4	10	155	234	257
7/8	9	206	372	382
1	8	310	551	587
1-1/8	7	480	872	794
1-1/4	7	675	1211	1105
1-3/8	6	900	1624	1500
1-1/2	6	1100	1943	1775
1-5/8	5.5	1470	2660	2425
1-3/4	5	1900	3436	3150
1-7/8	5	2360	4695	4200
2	4.5	2750	5427	4550

In order to determine the torque for a <u>fine thread</u> bolt, increase the above coarse thread ratings by 9%.

EFFECT OF LUBRICATION ON TORQUE

Lubricant	Torque Rating in Foot-Pounds	
	5/16-18 thread/inch	1/2-13 thread/inch
NO LUBE, steel	29	121
Plated & cleaned	19 (34%)	90 (26%)
SAE 20 oil	18 (38%)	87 (28%)
SAE 40 oil	17 (41%)	83 (31%)
Plated & SAE 30	16 (45%)	79 (35%)
White grease	16 (45%)	79 (35%)
Dry Mily film	14 (52%)	66 (45%)
Graphite & oil	13 (55%)	62 (49%)

Use the above lubrication percentages to calculate the approximate decrease in torque rating for other bolt sizes.

BOLT TORQUE SPECIFICATIONS

Bolt Size Inches	Coarse Thread /Inch	SAE Grade 6 133,000 psi Med. Carbon Temp. Steel	SAE Grade 7 133,000 psi Med. Carbon Alloy Steel	SAE Grade 8 150,000 psi Med. Carbon Alloy Steel
		Standard Dry Torque in Foot-Pounds		
1/4	20	12.5	13	14
5/16	18	24	25	29
3/8	16	43	44	47
7/16	14	69	71	78
1/2	13	106	110	119
9/16	12	150	154	169
5/8	11	209	215	230
3/4	10	350	360	380
7/8	9	550	570	600
1	8	825	840	700
1-1/8	7	1304	1325	1430
1-1/4	7	1815	1825	1975
1-3/8	6	2434	2500	2650
1-1/2	6	2913	3000	3200
1-5/8	5.5	3985	4000	4400
1-3/4	5	5189	5300	5650
1-7/8	5	6980	7000	7600
2	4.5	7491	7500	8200

In order to determine the torque for a fine thread bolt, increase the above coarse thread ratings by 9%.

Grades over Grade 8 are not common commercially, except in aircraft use. The following are a few of those types:

Supertanium, 160,000 psi, 8 points on head, quenched and tempered special alloy steel.

A354BD;A490, 150,000 psi, no markings, med. carbon quenched and tempered steel.

N.A.S. 144, MS2000, Military and Aircraft Std, 160,000 psi, high carbon alloy, quenched and tempered.

N.A.S. 623, National Aircraft Standard, 180,000 psi, high carbon alloy, quenched and tempered.

Aircraft Assigned Steel, no number, 220,000 psi, high carbon alloy, quenched and tempered.

Torque ratings for the above special alloy bolts should be obtained from the manufacturer.

BOLT TORQUE SPECIFICATIONS

Bolt Size Inches or Numbers	Coarse Thread /Inch	Allen Head 160,000 psi High Carbon CaseH Steel	Machine Scr. 60,000 psi Yellow Brass	Machine Scr. 70,000 psi Silicone Bronze
		Standard Dry Torque in Foot-Pounds		
#2	56	...	2 in#	2.3 in#
#3	48	...	3.3 in#	3.7 in#
#4	40	...	4.4 in#	4.9 in#
#5	40	...	6.4 in#	7.2 in#
#6	32	21	8 in#	10 in#
#8	32	46	16 in#	19 in#
#10	24	60	20 in#	22 in#
1/4	20	16	65 in#	70 in#
5/16	18	33	110 in#	125 in#
3/8	16	54	17	20
7/16	14	84	27	30
1/2	13	125	37	41
9/16	12	180	49	53
5/8	11	250	78	88
3/4	10	400	104	117
7/8	9	640	160	180
1	8	970	215	250
1-1/8	7	1520	325	365
1-1/4	7	2130	400	450
1-3/8	6	2850	...	...
1-1/2	6	3450	595	655
1-5/8		4700	...	...
1-3/4	5	6100	...	...
1-7/8		8200	...	...
2	4-1/2	8800	...	...

In order to determine the torque for a <u>fine thread</u> bolt, increase the above coarse thread ratings by 9%.

Socket Set Screws (looks like an allen head without the head) are usually rated at 212,000 psi, and are high carbon, case hardened steel. Torque ratings are as follows: #6 = 9 in-lbs, #8 = 16 in-lbs, #10 = 30 in-lbs, 1/4 in = 70 in-lbs, 5/16 in = 140 in-lbs, 3/8 in = 18 ft-lbs, 7/16 in = 29 ft-lbs, 1/2 in = 43 ft-lbs, 9/16 in = 63 ft-lbs, 5/8 in = 100 ft-lbs, and 3/4 in = 146 ft-lbs.

BOLT TORQUE SPECIFICATIONS

METRIC		5D	8G	10K
Bolt Size Milli-meters	Coarse Thread Pitch	Standard 5D 71,160 psi Med. Carbon Steel	Standard 8G 113,800 psi Med. Carbon Steel	Standard 10K 142,000 psi Med. Carbon Steel
		Standard Dry Torque in Foot-Pounds		
6 mm	1.00	5	6	8
8 mm	1.00	10	16	22
10 mm	1.25	19	31	40
12 mm	1.25	34	54	70
14 mm	1.25	55	89	117
16 mm	2.00	83	132	175
18 mm	2.00	111	182	236
22 mm	2.50	182	284	394
24 mm	3.00	261	419	570

METRIC		12K		
Bolt Size Milli-meters	Coarse Thread Pitch	Standard 12K 170,674 psi Med. Carbon Steel		
		Standard Dry Torque in Foot-Pounds		
6 mm	1.00	10		
8 mm	1.00	27		
10 mm	1.25	49		
12 mm	1.25	86		
14 mm	1.25	137		
16 mm	2.00	208		
18 mm	2.00	183		
22 mm	2.50	464		
24 mm	3.00	689		

In order to determine the torque for a fine thread bolt, increase the above coarse thread ratings by 9%. See first page of Bolt Torque Specifications for information on the Effects of Lubrication on Bolt Torque.

BOLT TORQUE SPECIFICATIONS

Whitworth		Grades A & B 62,720 psi Med. Carbon Steel	Grade S 112,000 psi Med. Carbon Steel	Grade T 123,200 psi Med. Carbon Steel
Bolt Size Inches	Coarse Thread /Inch			
		Standard Dry Torque in Foot-Pounds		
1/4	20	5	7	9
5/16	18	9	15	18
3/8	16	15	27	31
7/16	14	24	43	51
1/2	12	36	64	79
9/16	12	52	94	111
5/8	11	73	128	155
3/4	11	118	213	259
7/8	9	186	322	407
1	8	276	497	611

Whitworth		Grade V 145,600 psi Med. Carbon Steel		
Bolt Size Inches	Coarse Thread /Inch			
		Standard Dry Torque in Foot-Pounds		
1/4	20	10		
5/16	18	21		
3/8	16	36		
7/16	14	58		
1/2	12	89		
9/16	12	128		
5/8	11	175		
3/4	11	287		
7/8	9	459		
1	8	693		

In order to determine the torque for a <u>fine thread</u> bolt, increase the above coarse thread ratings by 9%. See first page of Bolt Torque Specifications for information on the Effects of Lubrication on Bolt Torque.

WOOD SCREW SPECIFICATIONS

Screw Number	Pilot Hole Sizes		Shank Diameter Inches	Shank Hole Clearance Drill Number
	Hard Wood Drill Number	Soft Wood Drill Number		
0	66	75	0.060	52
1	57	71	0.073	47
2	54	65	0.086	42
3	53	58	0.099	37
4	51	55	0.112	32
5	47	53	0.125	30
6	44	52	0.138	27
7	39	51	0.151	22
8	35	48	0.164	18
9	33	45	0.177	14
10	31	43	0.190	10
11	29	40	0.203	4
12	25	38	0.216	2
14	14	32	0.242	D
16	10	29	0.268	I
18	6	26	0.294	N
20	3	19	0.320	P
24	D	15	0.372	V

WOOD SCREW NUMBER vs STD LENGTHS

Screw Number	Standard Lengths in Inches
0	1/4
1	1/4, 3/8
2	1/4, 3/8, 1/2
3	1/4, 3/8, 1/2, 5/8
4	3/8, 1/2, 5/8, 3/4
5	3/8, 1/2, 5/8, 3/4
6	3/8, 1/2, 5/8, 3/4, 7/8, 1, 1-1/4, 1-1/2
7	3/8, 1/2, 5/8, 3/4, 7/8, 1, 1-1/4, 1-1/2
8	1/2, 5/8, 3/4, 7/8, 1, 1-1/4, 1-1/2, 1-3/4, 2
9	5/8, 3/4, 7/8, 1, 1-1/4, 1-1/2, 1-3/4, 2, 2-1/4
10	5/8, 3/4, 7/8, 1, 1-1/4, 1-1/2, 1-3/4, 2, 2-1/4, 2-1/2
11	3/4, 7/8, 1, 1-1/4, 1-1/2, 1-3/4, 2, 2-1/4, 2-1/2, 3
12	7/8, 1, 1-1/4, 1-1/2, 1-3/4, 2, 2-1/4, 2-1/2, 2-3/4, 3, 3-1/2
14	1, 1-1/4, 1-1/2, 1-3/4, 2, 2-1/4, 2-1/2, 2-3/4, 3, 3-1/2, 4, 4-1/2
16	1-1/4, 1-1/2, 1-3/4, 2, 2-1/4, 2-1/2, 2-3/4, 3, 3-1/2, 4, 4-1/2, 5, 5-1/2
18	1-1/2, 1-3/4, 2, 2-1/4, 2-1/2, 2-3/4, 3, 3-1/2, 4, 4-1/2, 5, 5-1/2, 6
20	1-3/4, 2, 2-1/4, 2-1/2, 2-3/4, 3, 3-1/2, 4, 4-1/2, 5, 5-1/2, 6
24	3-1/2, 4, 4-1/2, 5, 5-1/2, 6

SHEET METAL SCREW SPECS

Screw Diameter # (inch)	Thickness of Metal Gauge #	Diameter of Pierced Hole (inch)	Drilled Hole Size Drill Number
#4 (0.112)	28	0.086	44
	26	0.086	44
	24	0.093	42
	22	0.098	42
	20	0.100	40
#6 (0.138)	28	0.111	39
	26	0.111	39
	24	0.111	39
	22	0.111	38
	20	0.111	36
#7 (0.155)	28	0.121	37
	26	0.121	37
	24	0.121	35
	22	0.121	33
	20	0.121	32
	18	....	31
#8 (0.165)	26	0.137	33
	24	0.137	33
	22	0.137	32
	20	0.137	31
	18	....	30
#10 (0.191)	26	0.158	30
	24	0.158	30
	22	0.158	30
	20	0.158	29
	18	0.158	25
#12 (0.218)	24	....	26
	22	0.185	25
	20	0.185	24
	18	0.185	22
#14 (0.251)	24	....	15
	22	0.212	12
	20	0.212	11
	18	0.212	9

Note: The above values are recommended average values only. Variations in materials and local conditions may require significant deviations from the recommended values.

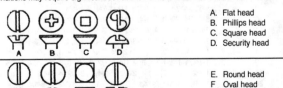

A. Flat head
B. Phillips head
C. Square head
D. Security head

E. Round head
F. Oval head
G. Square head
H. Pan head

CABLE CLAMPS FOR WIRE ROPE

Rope Diameter Inches	Number of Clamps Required	Clip Spacing Inches	Rope Turn-back Inches
1/8	2	3	3-1/4
3/16	2	3	3-3/4
1/4	2	3-1/4	4-3/4
5/16	2	3-1/4	5-1/4
3/8	2	4	6-1/2
7/16	2	4-1/2	4
1/2	3	5	11-1/2
9/16	3	5-1/2	12
5/8	3	5-3/4	12
3/4	4	6-3/4	18
7/8	4	8	19
1	5	8-3/4	26
1-1/8	6	9-3/4	34
1-1/4	6	10-3/4	37
1-7/16	7	11-1/2	44
1-1/2	7	12-1/2	48
1-5/8	7	13-1/4	51
1-3/4	7	14-1/2	53
2	8	16-1/2	71
2-1/4	8	16-1/2	73
2-1/2	9	17-3/4	84
2-3/4	10	18	100
3	10	18	106

AMERICAN NATIONAL TAPS & DIES

	Fine Threads			Coarse Threads		
Thread	Threads / Inch	Tap Drill	Tap Decimal Inch	Threads / Inch	Tap Drill	Tap Decimal Inch
#0	80	3/64	0.0469	...	...	...
#1	72	#53	0.0595	64	#53	0.0595
#2	64	#50	0.0700	56	#50	0.0700
#3	56	#45	0.0820	48	#47	0.0785
#4	48	#42	0.0935	40	#43	0.0890
1/8	40	#38	0.1015	32	3/32	0.0938
#5	44	#37	0.1040	40	#38	0.1015
#6	40	#33	0.1130	32	#36	0.1065
#8	36	#29	0.1360	32	#29	0.1360
3/16	32	#22	0.1570	24	#26	0.1470
#10	32	#21	0.1590	24	#25	0.1495
#12	28	#14	0.1820	24	#16	0.1770
1/4	28	#3	0.2130	20	#7	0.2010
5/16	24	I	0.2720	18	F	0.2570
3/8	24	Q	0.3320	16	5/16	0.3125
7/16	20	25/64	0.3906	14	U	0.3680
1/2	20	29/64	0.4531	13	27/64	0.4219
9/16	18	33/64	0.5156	12	31/64	0.4844
5/8	18	37/64	0.5781	11	17/32	0.5313
3/4	16	11/16	0.6875	10	21/32	0.6563
7/8	14	13/16	0.8125	9	49/64	0.7656
1	14	15/16	0.9375	8	7/8	0.8750
1-1/8	12	1-3/64	1.0469	7	63/64	0.9844
1-1/4	12	1-11/64	1.1719	7	1-7/64	1.1094
1-3/8	12	1-19/64	1.2969	6	1-7/32	1.2188
1-1/2	12	1-27/64	1.4219	6	1-21/64	1.3281
1-3/4	...	...	...	2	1-35/64	1.5469
2	...	...	...	4-1/2	1-25/32	1.7813
2-1/4	...	...	...	4-1/2	2-1/32	2.0313
2-1/2	...	...	...	4	2-1/4	2.2500
2-3/4	...	...	...	4	2-1/2	2.5000
3	...	...	...	4	2-3/4	2.7500
3-1/4	...	...	...	4	3	3.0000
3-1/2	...	...	...	4	3-1/4	3.2500
3-3/4	...	...	...	4	3-1/2	3.5000
4	...	...	...	4	3-3/4	3.7500

Note: That there are other sizes and thread per inch taps and dies available, e.g. 1/4 inch can have 10,12,14,16,18,22,23,24,25,26,27,30,32,34,36,38,40,42,44,48,50,52,56,60,64,72, and 80 threads/inch depending on the standard. Threads shown in the table above are simply the most common.

METRIC TAPS AND DIES

Fine Threads		Coarse Threads		Tap Drill Size	
mm	Inches	French	International	mm	inches
1.5	0.0590	0.35	...	1.10	0.0433
2	0.0787	0.45	...	1.50	0.0590
2	0.0787	...	0.40	1.60	0.0630
2.3	0.0895	...	0.40	1.90	0.0748
2.5	0.0984	0.45	...	2.00	0.0787
2.6	0.1124	...	0.45	2.10	0.0827
3	0.1181	...	0.5	2.50	0.0984
3	0.1181	0.60	...	2.40	0.0945
3.5	0.1378	0.60	0.60	2.90	0.1142
4	0.1575	0.75	...	3.25	0.1279
4	0.1575	...	0.70	3.30	0.1299
4.5	0.1772	0.75	0.75	3.75	0.1476
5	0.1968	0.90	...	4.10	0.1614
5	0.1968	...	0.80	4.20	0.1653
5.5	0.2165	0.90	0.90	4.60	0.1811
6	0.2362	1.00	1.00	5.00	0.1968
7	0.2856	1.00	1.00	6.00	0.2362
8	0.3150	1.00	...	7.00	0.2756
8	0.3150	...	1.25	6.80	0.2677
9	0.3543	1.00	...	8.00	0.3150
9	0.3543	...	1.25	7.80	0.3071
10	0.3937	1.50	1.50	8.60	0.3386
11	0.3937	...	1.50	9.60	0.3780
12	0.4624	1.50	...	10.50	0.4134
12	0.4624	...	1.75	10.50	0.4134
14	0.5512	2.00	2.00	12.00	0.4724
16	0.6299	2.00	2.00	14.00	0.5118
18	0.7087	2.50	2.50	15.50	0.6102
20	0.7974	2.50	2.50	17.50	0.6890
22	0.8771	2.50	2.50	19.50	0.7677
24	0.9449	3.00	3.00	21.00	0.8268
26	1.0236	3.00	...	23.00	0.9055
27	1.0630	...	3.00	24.00	0.9449
28	1.1024	3.00	...	25.00	0.9842
30	1.1811	3.50	3.50	26.50	1.0433
32	1.2598	3.50	...	28.50	1.1220
33	1.2992	...	3.50	29.50	1.1614
34	1.3386	3.50	...	30.50	1.2008
36	1.4173	4.00	4.00	32.00	1.2598
38	1.4961	4.00	...	34.00	1.3386
39	1.5354	...	4.00	35.00	1.3779
40	1.5748	4.00	...	36.00	1.4173
42	1.6535	4.50	4.50	37.00	1.4567

DRILL & CUTTING LUBRICANTS

Material to be Worked	Machine Process		
	Drilling	Threading	Lathe
Aluminum	Soluble oil Kerosene Lard oil	Soluble oil, Kerosene, & Lard oil	Soluble oil
Brass	Dry Soluble oil Kerosene Lard oil	Soluble oil Lard oil	Soluble oil
Bronze	Dry Soluble oil Mineral oil Lard oil	Soluble oil Lard oil	Soluble oil
Cast Iron	Dry Air jet Soluble oil	Dry Sulphurized oil Mineral lard oil	Dry Soluble oil
Copper	Dry Soluble oil Mineral lard oil Kerosene	Lard oil Lard oil	Soluble oil
Malleable iron	Dry Soda water	Lard oil Soda water	Soluble oil Soda water
Monel metal	Soluble oil Lard oil	Lard oil	Soluble oil
Steel alloys	Soluble oil Sulphurized oil Mineral lard oil	Sulphurized oil Lard oil	Soluble oil
Steel, machine	Soluble oil Sulphurized oil Lard oil Mineral lard oil	Soluble oil Mineral lard oil	Soluble oil
Steel, tool	Soluble oil Sulpherized oil Mineral lard oil	Sulpherized oil Lard oil	Soluble oil

DRILLING SPEEDS vs MATERIAL

Material	Speed rpm	Description
Cast iron	6000 to 6500	1/16 inch drill
	3500 to 4500	1/8 inch drill
	2500 to 3000	3/16 inch drill
	2000 to 2500	1/4 inch drill
	1500 to 2000	5/16 inch drill
	1500 to 2000	3/8 inch drill
	1000 to 1500	> 7/16 inch drill
Glass	700	Special metal tube drilling
Plastics	6000 to 6500	1/16 inch drill
	5000 to 6000	1/8 inch drill
	3500 to 4000	3/16 inch drill
	3000 to 3500	1/4 inch drill
	2000 to 2500	5/16 inch drill
	1500 to 2000	3/8 inch drill
	500 to 1000	> 7/16 inch drill
Soft Metals (copper)	6000 to 6500	1/16 inch drill
	6000 to 6500	1/8 inch drill
	5000 to 6000	3/16 inch drill
	4500 to 5000	1/4 inch drill
	3500 to 4000	5/16 inch drill
	3000 to 3500	3/8 inch drill
	1500 to 2500	> 7/16 inch drill
Steel	5000 to 6500	1/16 inch drill
	3000 to 4000	1/8 inch drill
	2000 to 2500	3/16 inch drill
	1500 to 2000	1/4 inch drill
	1000 to 1500	5/16 inch drill
	1000 to 1500	3/8 inch drill
	500 to 1000	> 7/16 inch drill
Wood	4000 to 6000	Carving and routing
	3800 to 4000	All woods, 0 to 1/4 inch drills
	3100 to 3800	All woods, 1/4 to 1/2 inch drills
	2300 to 3100	All woods, 1/2 to 3/4 inch drills
	2000 to 2300	All woods, 3/4 to 1 inch drills
	700 to 2000	All woods, > 1 inch drills, fly cutters, and multi-spur bits
	< 700	

GAS WELDING RODS

Rod Diameter	Rods per Pound (36 inch long)			
	Steel	Brass	Aluminum	Cast Iron
1/16	.31	29	91	NA
3/32	.14	13	41	NA
1/8	.8	7	23	NA
5/32	.5	NA	NA	NA
3/16	.3-1/2	3	9	5-1/2
1/4	.2	2	6	2-1/4
5/16	.1-1/3	NA	NA	1/2
3/8	.1	1	NA	1/4

WELDING GASES

Gas	Tank Sizes Cubic Ft	Comments
Acetylene	300 100 75 40 10	Formula - C_2H_2, explosive Colorless, flammable gas, garlic-like odor, explosion danger if used in welding with gage pressures over 15 psig (30 psig absolute).
Oxygen	244,122 80,40,20 4500 liquid	Formula - O_2, non-explosive Colorless, odorless, tasteless. Supports combustion in welding.
Nitrogen	225,113 80,40,20	Formula - N_2, non-explosive Colorless, odorless, tasteless, inert.
Argon	330, 131 4754 liquid	Formula - Ar, non-explosive Colorless, odorless, tasteless, inert.
Carbon Dioxide	50 lbs 20 lbs	Formula - CO_2, non-explosive Toxic in large quantities Colorless, odorless, tasteless, inert.
Hydrogen	191	Formula - H_2, explosive Colorless, odorless, tasteless Lightest gas known.
Helium	221	Formula - He, non-explosive Colorless, odorless, tasteless, inert.

HARD & SOFT SOLDER ALLOYS

Metal to be Soldered	Tin	Lead	Zinc	Copper	Other
SOFT SOLDER:					
Aluminum	70		25		Al=3, Pho=2
Bismuth	33	33			Bi=34
Block tin	99	1			
Brass	66	34			
Copper	60	40			
Gold	67	33			
Gun metal	63	37			
Iron & Steel	50	50			
Lead	33	67			
Pewter	25	25			Bi=50
Silver	67	33			
Steel, galvanized	58	42			
Steel, tinned	64	36			
Zinc	55	45			
HARD SOLDER:					
Brass, soft			78	22	
Brass, hard			55	45	
Copper			50	50	
Iron, Cast			45	55	
Iron & Steel			36	64	
Gold				22	Ag=11, Au=67
Silver			10	20	Ag=70

FLUX	Metal to be used on
Ammonia Chloride	Galvanized iron, iron, nickel, tin, zinc, brass, copper, gun metal
Borax	For hard solders, brass, copper, gold, iron & steel, silver
Cuprous oxide	Cast iron
Hydrochloric Acid	Galvanized iron and steel, tin, zinc
Organic	Lead, pewter
Resin	Brass, bronze, cadmium, copper, lead, silver, gun metal, tinned steel
Stainless Steel Flux	Special for stainless steel only
Sterling	Silver
Tallow	Lead, pewter
Zinc Chloride	Bismuth, tin, brass, copper, gold, silver, gun metal, tinned steel

FIRE EXTINGUISHERS

Fire extinguishers are an absolute must in any shop, garage, home, automobile, or business. Fire extinguishers are classified by the types of fires they will put out and the size of the fire they will put out. The basic types are as follows:

TYPE A: For wood, cloth, paper, trash and other common materials. These fires are put out by "heat absorbing" water or water based materials or smothered by dry chemicals.

TYPE B: For oil, gasoline, grease, paints & other flammable liquids. These fires are put out by smothering, preventing the release of combustible vapors, or stopping the combustion chain. Use Halon, dry chemicals, carbon dioxide, or foam.

TYPE C: For "live" electrical equipment. These fires are put out by the same process as TYPE B, but the extinguishing material <u>must be electrically non-conductive.</u> Use halon, dry chemicals, or carbon dioxide.

TYPE D: For combustible metals such as magnesium. These fires must be put out by heat absorption and smothering. Obtain specific information on these requirements from the fire department.

Combinations of the above letters indicate the extinguisher will put out more than one type of fire. For example, a Type ABC unit will put out all three types of fires. The "size" of the fire an extinguisher will put out is shown by a number in front of the Type, such as "10B". The base line numbers are as follows:

Class "1A": Will put out a stack of 50 burning sticks that are 20 inches long each.

Class "1B": Will put out an area of burning naptha that is 2.5 square feet in size.

Any number other than the "1" simply indicates the extinguisher will put out a fire that many times larger, for example "10A" will put out a fire 10 times larger than "1A".

Some general recommendations when purchasing a fire extinguisher are as follows:

1: Buy TYPE ABC so that you never have to think about what type of fire you are using it on. Also, buy a halon or carbon dioxide extinguisher so you don't damage electronic equipment and there is much less mess. The relative prices of extinguishers are Foam (very expensive - used on big fires such as aircraft), Carbon Dioxide (also expensive but leaves no mess), Halon (medium range prices), and Dry Chemical (very inexpensive, but leaves a mess).

2: Buy units with metal components and a gauge and are approved by Underwriters Labs or other testing group. Plastic units are generally poorly constructed and break easily; buy good extinguishers, your life and property may depend on it!

3: Buy more than one extinguisher and mount them on the wall near escape routes so that children can reach them.

4: Study the instructions when you get the unit, there may not be time after a fire has started.

PULLEYS AND GEARS

For single reduction or increase of speed by means of belting where the speed at which each shaft should run is known, and one pulley is in place:

> Multiply the diameter of the pulley which you have by the number of revolutions per minute that its shaft makes; divide this product by the speed in revolutions per minute at which the second shaft should run. The result is the diameter of pulley to use.

Where both shafts with pulleys are in operation and the speed of one is known:

> Multiply the speed of the shaft by diameter of its pulley and divide this product by diameter of pulley on the other shaft. The result is the speed of the second shaft.

Where a countershaft is used, to obtain size of main driving or driven pulley, or speed of main driving or driven shaft, it is necessary to calculate, as above, between the known end of the transmission and the countershaft, then repeat this calculation between the countershaft and the unknown end.

A set of gears of the same pitch transmits speeds in proportion to the number of teeth they contain. Count the number of teeth in the gear wheel and use this quantity instead of the diameter of pulley, mentioned above, to obtain number of teeth cut in unknown gear, or speed of second shaft.

Rule for Finding Size of Pulleys

$$d = \frac{D \times S}{S'} \qquad\qquad D = \frac{d \times S'}{S}$$

d = diameter of driven pulley.

D = diameter of driving pulley.

S = number of revolutions per minute of driving pulley.

S' = number of revolutions per minute of driven pulley.

PULLEYS AND GEARS (CONT.)

Rule for Finding Size of Gears

$$n = \frac{N \times S}{S'} \qquad N = \frac{n \times S'}{S}$$

n = number of teeth in pinion (driving gear).

N = number of teeth in gear (driven gear).

S = number of revolutions per minute of pinion.

S' = number of revolutions per minute of gear.

Shafting

The formula for determining the size of steel shaft for transmitting a given power at a given speed is as follows:

$$\text{diameter of shaft in inches} = \sqrt[3]{\frac{K \times HP}{RPM}}$$

when HP = the horse power to be transmitted

RPM = speed of shaft

K = factor which varies from 50 to 125 depending on type of shaft and distance between supporting bearings.

For line shaft having bearings 8 feet apart:

K = 90 for turned shafting

K = 70 for cold-rolled shafting

Belts

The following formula is used to determine the length of belting:

$$\text{length of belt} = \frac{3.14 (D + d)}{2} + 2 \sqrt{X^2 + \left(\frac{D - d}{2}\right)^2}$$

when D = diameter of large pulley

d = diameter of small pulley

X = distance between centers of shafting

STANDARD "V" BELT LENGTHS

A BELTS			B BELTS			C BELTS		
BELT # Standard	Pitch Length	Outside Length	BELT # Standard	Pitch Length	Outside Length	BELT # Standard	Pitch Length	Outside Length
A26	27.3	28.0	B35	36.8	38.0	C51	53.9	55.0
A31	32.3	33.0	B38	39.8	41.0	C60	62.9	64.0
A35	36.3	37.0	B42	43.8	45.0	C68	70.9	81.0
A38	39.3	40.0	B46	47.8	49.0	C75	77.9	79.0
A42	43.3	44.0	B51	52.8	54.0	C81	83.9	85.0
A46	47.3	48.0	B55	56.8	58.0	C85	87.9	89.0
A51	52.3	53.0	B60	61.8	63.0	C90	92.9	94.0
A55	56.3	57.0	B68	69.8	71.0	C96	98.9	100.0
A60	61.3	62.0	B75	76.8	78.0	C105	107.9	109.0
A68	69.3	70.0	B81	82.8	84.0	C112	114.9	116.0
A75	76.3	77.0	B85	86.8	88.0	C120	122.9	124.0
A80	81.3	82.0	B90	91.8	93.0	C128	130.9	132.0
A85	86.3	87.0	B97	98.8	100.0	C136	138.9	140.0
A90	91.3	92.0	B105	106.8	108.0	C144	146.9	148.0
A96	97.3	98.0	B112	113.8	115.0	C158	160.9	162.0
A105	106.3	107.0	B120	121.8	123.0	C162	164.9	166.0
A112	113.3	114.0	B128	129.8	131.0	C173	175.9	177.0
A120	121.3	122.0	B136	137.8	139.0	C180	182.9	184.0
A128	129.3	130.0	B144	145.8	147.0	C195	197.9	199.0
			B158	159.8	161.0	C210	212.9	214.0
			B173	174.8	176.0	C240	240.9	242.0
			B180	181.8	183.0	C270	270.9	272.0
			B195	196.8	198.0	C300	300.9	302.0

D BELTS		
BELT # Standard	Pitch Length	Outside Length
D120	123.3	125.0
D128	131.3	133.0
D144	147.3	149.0
D158	1613	163.0
D162	165.3	167.0
D173	176.3	178.0
D180	183.3	185.0
D195	198.3	200.0
D210	213.3	215.0
D240	240.8	242.0
D270	270.8	272.5
D300	300.8	302.5
D330	330.8	332.5
D360	360.8	362.5
D390	390.8	392.5
D420	420.8	422.5
D480	480.8	482.5
D540	540.8	542.5
D600	600.8	602.5

(B BELTS continued)

BELT # Standard	Pitch Length	Outside Length
B210	211.8	213.0
B240	240.3	241.5
B270	270.3	271.5
B300	300.3	301.5

(C BELTS continued)

BELT # Standard	Pitch Length	Outside Length
C360	360.9	362.0
C390	390.9	392.0
C420	420.9	422.0

3V BELTS		5V BELTS		8V BELTS	
Belt No.	Belt Length	Belt No.	Belt Length	Belt No.	Belt Length
3V250	25.0	5V500	50.0	8V1000	100.0
3V265	26.5	5V530	53.0	8V1060	106.0
3V280	28.0	5V560	56.0	8V1120	112.0
3V300	30.0	5V600	60.0	8V1180	118.0
3V315	31.5	5V630	63.0	8V1250	125.0
3V335	33.5	5V670	67.0	8V1320	132.0
3V355	35.5	5V710	71.0	8V1400	140.0
3V375	37.5	5V750	75.0	8V1500	150.0
3V400	40.0	5V800	80.0	8V1600	160.0

STANDARD "V" BELT LENGTHS (CONT.)

E BELTS		
BELT # Standard	Pitch Length	Outside Length
E180	184.5	187.5
E195	199.5	202.5
E210	214.5	217.5
E240	241.0	244.0
E270	271.0	274.0
E300	301.0	304.0
E330	331.0	334.0
E360	361.0	364.0
E390	391.0	394.0
E420	421.0	424.0
E480	481.0	484.0
E540	541.0	544.0
E600	601.0	604.0

3V BELTS		5V BELTS		8V BELTS	
3V425	42.5	5V850	85.0	8V1700	170.0
3V450	45.0	5V900	90.0	8V1800	180.0
3V475	47.5	5V950	95.0	8V1900	190.0
3V500	50.0	5V1000	100.0	8V2000	200.0
3V530	53.0	5V1060	106.0	8V2120	212.0
3V560	56.0	5V1120	112.0	8V2240	224.0
3V600	60.0	5V1180	118.0	8V2360	236.0
3V630	63.0	5V1250	125.0	8V2500	250.0
3V670	67.0	5V1320	132.0	8V2650	265.0
3V710	71.0	5V1400	140.0	8V2800	280.0
3V750	75.0	5V1500	150.0	8V3000	300.0
3V800	80.0	5V1600	160.0	8V3150	315.0
3V850	85.0	5V1700	170.0	8V3350	335.0
3V900	90.0	5V1800	180.0	8V3550	355.0
3V950	95.0	5V1900	190.0	8V3750	375.0
3V1000	100.0	5V2000	200.0	8V4000	400.0
3V1060	106.0	5V2120	212.0	8V4250	425.0
3V1120	112.0	5V2240	224.0	8V4500	450.0
3V1180	118.0	5V2360	236.0	*8V5000	500.0
3V1250	128.0	5V2500	250.0		
3V1320	132.0	5V2650	265.0		
3V1400	140.0	5V2800	280.0		
		5V3000	300.0		
		5V3150	315.0		
		5V3350	335.0		
		5V3550	355.0		

For example, if the 60-inch "B" section belt shown is manufactured 3/10 of an inch longer, it will be code marked 53 rather than the 50 shown. If made 3/10 of an inch shorter, it will be code marked 47. While both of these belts have the belt number B60 they cannot be used satisfactorily in a set because of the difference in their actual length.

TYPICAL CODE MARKING

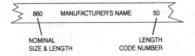

860	MANUFACTURER'S NAME	50
NOMINAL SIZE & LENGTH		LENGTH CODE NUMBER

CHAPTER 12
TOLL-FREE NUMBERS FOR ELECTRICAL SUPPLIERS

Electric Equipment - Manufacturer's

800-345-1280	AESI/Stedi Watt Pure Power Systems (Chambersburg PA)
800-343-6434	Aemc Instruments (Dover NE)
800-233-4717	Arlington Industries (Scranton PA)
800-428-8637	Automatic Doorman (Palisades Park NJ)
800-233-5225	Blackjack Electric (Winter Springs FL)
800-821-8201	Challenger Electrical Equip-Fax Line (Orlando FL)
800-343-1391	Chauvin Arnoux, Inc (Boston MA)
800-472-0717	Con J Franke Electric Inc (Stockton CA)
800-266-0908	Consolidated Electrical Distributors (Long Beach CA)
800-683-1796	Consolidated Electrical Distributors (Harlingen TX)
800-814-1613	Ditch Witch of South Carolina (Lexington SC)
800-858-4751	Electric Busway Corp (Bethel Park PA)
800-331-4121	Environmental Solutions (Tampa FL)
800-828-0920	G & S International Corp (Pine Brook NJ)
800-255-6993	IRP Professional Sound (Bensenville IL)
800-262-4332	Idec (Sunnyvale CA)
800-947-1110	JST Corp (Mt Prospect IL)
800-352-4563	KME DJA-Surface Mounting Technology (Elk Grove Village IL)
800-554-6096	Kirkwood Industries (Conway AR)
800-206-0035	Magnetek Lighting Inc-Customer Service (Nashville TN)
800-535-7878	Monnex Industries (Arlington Heights IL)
800-921-3600	Parts & Electric Motors Inc (Chicago IL)
800-331-3122	Platt Electric Supply Whsle (Beaverton OR)
800-382-3765	Relcom (Forest Grove OR)
800-237-8859	Satin American-Fax (Shelton CT)
800-272-7711	Satin American-Fax (Shelton CT)
800-798-1261	Stitzell Electric Supply (Des Moines IA)
800-548-7763	Ventron Corp (Galena OH)
800-800-7128	Video Messenger Co (Stamford CT)

| 800-838-1482 | Wadsley Electric (Alta IA) |
| 800-297-3535 | Zang Agency H C (Buffalo, NY) |

Electric Equipment & Supplies - Wholesale

800-523-9065	Breakers & Controls, Inc (Clifton NJ)
800-999-0055	Cameron & Barkley (Augusta GA)
800-524-7501	Circuit Supply & Service (San Dimas CA)
800-524-7502	Circuit Supply & Service-Fax Line (San Dimas CA)
800-523-5232	Electric Maintenance & Engineering Works (Woonsocket RI)
800-588-1094	Electric Supply Of Brownwood (Brownwood TX)
800-588-1556	Electric Supply of Seguin (Seguin TX)
800-562-5581	Electrical Equip Sales (Raonoke VA)
800-962-7964	Electrical Sales Agents (Syracuse NY)
800-925-8277	Exide Electronics (Oakbrook IL)
800-774-6489	Foster Electric Co., Inc (Lynchburg VA)
800-424-3111	General Magnetic Electric Wholesale (Commerce CA)
800-722-5635	Jo El Electric Supply (El Dorado AR)
800-999-0880	Langstadt Electric (Madison WI)
800-282-4771	Microwave Components (Stuart FL)
800-242-3143	Nelson Electric Supply (Racine WI)
800-428-0430	Platt Electric Supply (Concord CA)
800-747-4367	Rock Island Electric Motor Repair (Rock Island IL)
800-269-7290	Romac Supply Co (Los Angeles CA)
800-521-5092	San Diego Whsle Electric (San Diego CA)
800-255-4324	Utility Safety Sales Division (Townsend MA)
800-552-7777	Wabash Electric Supply, Inc (Wabash IN)
800-938-8237	Westrex (Atlanta GA)

Electric Equipment - Used

800-807-9579	Eagle Circuit Breaker & Control Inc (Norcross GA)
800-328-1842	Emsco (Upland CA)
800-424-3111	General Magnetic Electric Wholesale (Commerce CA)
800-541-9212	Hite (Somerset PA)
800-231-0097	Indiana Coil Service-Fax Line (South Bend IN)
800-738-0303	Raritan Electrical Supply Co Inc (Edison NJ)

Electric Instruments

800-358-5545	Davis Instruments (Baltimore MD)
800-562-7115	E P L Inc (Iron Mountain MI)
800-452-2344	Laser Labs (Rockland MA)
800-678-6121	Platt Electric Supply (Beaverton OR)
800-582-4545	Wabash Electric Supply (Bluffton IN)

Electric Motors & Controls

800-558-8411	A O Smith Corp Hdqrs (Milwaukee WI)
800-252-6011	Alamo Electrical Supply (Duncanville TX)
800-257-7474	American Electric Motors (Memphis TN)
800-446-0408	American Transformer Corp (S El Monte CA)
800-426-5913	Aseptico (Kirkland WA)
800-346-1637	B & B Electric Motors (Seattle WA)
800-845-1020	B & B Motor & Control Corp (Bala-Cynwyd PA)
800-364-7521	Balance Electric (Hot Springs AR)
800-322-3460	Bearings Inc (Cleveland OH)
800-537-7499	Beavers Electric Inc (Battle Creek MI)
800-466-0449	Brink Industrial Technologies (Logansport IN)
800-922-0602	Brownell Electro Inc (Edison NJ)
800-458-9702	Cannon Electric Motor (Cannon Falls MN)
800-232-1231	Cherokee Electric Motor Service (Gaffney SC)
800-638-9802	Cleveland Electric Coil (Birmingham AL)
800-448-5877	Cleveland Electric Motors Inc (Shelby NC)
800-521-9148	Dart Electric Motor & Supply (Hamtramck MI)
800-730-4362	E M C Technologies (Tulsa OK)
800-352-8968	Electric Motor & Bearing Service (Albert Lea MN)
800-456-0066	Electric Motor Repair (Sioux Falls SD)
800-564-3185	Electric Motor & Controls (Titusville FL)
800-440-8801	Electric Motor & Drives Inc (Anderson SC)
800-458-6566	Electric Regulator Corp (CA)
800-279-5524	Electrical Mechanical Service (ST Paul MN)
800-299-2955	Electro Mechanical Services Inc (Cocoa FL)
800-256-9970	Electro Motor Inc (Sulphur LA)
800-543-0734	Electro Motors (Decatur AL)
800-633-9788	Electronic Controls Inc (N Miami FL)
800-681-7306	Electronic Technical Services (Porter TX)
800-678-1985	Elektrin/Toolmex (Schaumburg IL)
800-633-6951	Eltech Control (Marietta GA)
800-396-6427	Exonic Systems (Pittsburgh PA)
800-932-8986	Eylander Electric (Everett WA)
800-443-7483	Fincor Electronics (York PA)
800-635-7576	G & G Service (Anderson SC)
800-362-7634	General Control Systems (San Dimas CA)
800-821-5937	General Machinery (Birmingham AL)
800-424-3111	General Magnetic Electric Wholesale (US)
800-323-1106	Glenn Electric Inc (Mokena IL)
800-854-9049	Global D C Motor Components Co (McComb MS)
800-444-9712	Graybar Electric (Cleveland OH)

800-240-3260	Green Electric Motor Service (Ocala FL)
800-521-2258	Greer Electric Motors (Greer SC)
800-344-6049	Hackworth Electric Motors Inc (Wooster OH)
800-233-0614	Hancock Electric (Lyons KS)
800-692-7316	Heim Electric Co Inc (Harrisburg PA)
800-422-7493	Herold & Mielenz (Sacramento CA)
800-422-5270	Holander Electric Inc (Attleboro MA)
800-346-9350	Howell Electric Motors (Plainfield NJ)
800-446-4808	Huntingdon Electric (Huntingdon PA)
800-345-2264	I V S Inc (Freeport NY)
800-682-6160	Idec Corp (Northborough MA)
800-327-7858	Industrial Commutator (Maryville TN)
800-462-3430	Industrial Motor & Control (Phoenix AZ)
800-237-3786	Infranor Inc (Naugatuck CT)
800-208-3008	Integra LLC (Longmont CO)
800-572-3022	J & H Electric Cp (Providence RI)
800-828-1772	Jasper Electric Service (Jasper TX)
800-223-4898	K & G Electric Motor & Pump (Wantagh NY)
800-221-8577	Ken Coil Inc (Belle Chasse LA)
800-776-3629	Kurz Electric Service Corp (Appleton WI)
800-858-0091	L & L Machinery Co Inc (Olympia WA)
800-572-8570	Lincoln Electric Co-Service Center (Flint MI)
800-572-8569	Lincoln Electric-Service Center (Dayton OH)
800-572-8572	Lincoln Service Center (Indianapolis IN)
800-572-8571	Lincoln-Delco Service Center (Ferndale MI)
800-456-5058	Ling Products (Neenah WI)
800-622-0069	M & M Forklift Motor Repairs (St Louis MO)
800-544-8303	M N E Enterprises Inc (Bridge City TX)
800-325-7344	Magnetek (St Louis MO)
800-468-2062	Magnetek (La Vergne TN)
800-672-6495	Magnetek (La Vergne TN)
800-356-6266	Mann Electric Repair (Columbia SC)
800-262-3623	Manolo Garcia Electric Motors (Miami FL)
800-223-9280	Marion Electric Motor Service Inc (Marion OH)
800-843-2211	Midcontinent Rebuilders (Wichita KS)
800-258-5258	Mohler Armature & Electric (Boonville IN)
800-542-4077	Motector Systems (Fairfax VA)
800-245-0240	Motor Coils (Pittsburgh PA)
800-248-3841	Motor Products (Owosso MI)
800-871-8969	Motor Shop The (Gastonia NC)
800-428-5231	Motor Technology (Syracuse NY)
800-632-9060	Motor Technology (York PA)

800-252-2038	National Electric Coil Magnetek (Tucson AZ)
800-456-4264	Northwest Electric (Columbus NE)
800-234-2754	Northwest Electric (Kearney NE)
800-243-0520	Olsen Machine & Tool (N Abington MA)
800-309-7999	Oriental Motor USA-Fax (Torrance CA)
800-554-7005	Parts & Electric Motors (Chicago IL)
800-921-1400	Parts & Electric Motors Inc (Chicago IL)
800-354-6767	Power Systems (Boise ID)
800-242-2117	Precision Devices Inc (Wallingford CT)
800-878-7095	Premco (New Rochelle NY)
800-783-4321	R & R Electric (Columbus OH)
800-323-7049	RAE (McHenry IL)
800-835-3730	Rainbow Electric (Franklin Park IL)
800-269-7290	Romac Supply Co (Los Angeles CA)
800-669-9189	Safe-Way Electric Motor Company Inc (Providence RI)
800-707-7267	Sam's Industrial Electric Motors (Ponca City OK)
800-253-5172	Sanders Electric Motor Service (Lenoir NC)
800-452-3811	Schoen's Motors (Albany OR)
800-435-4066	Siemens Industrial Controls Distributor (Joliet IL)
800-426-6231	Smith Gray Electric (Columbus GA)
800-468-0160	Southern Electric Motor Sales & Service (Russellville KY)
800-795-0698	Steve S Electric (Los Angeles)
800-874-5909	Stevens Electric (Memphis TN)
800-834-8032	Stroudsburg Electric Motor Service (Stroudsburg PA)
800-782-9675	Sun Star Electric Inc (Lubbock TX)
800-628-1813	Superior Motor Winding (Cedar Rapids IA)
800-862-8813	Sytek Alsay (Houston TX)
800-553-3525	Sytek (Blacklick OH)
800-447-8962	Sytek (Cincinnati OH)
800-637-9835	Sytek-Benkiser (Sacramento CA)
800-513-7931	T & N Electric (Bluefield VA)
800-227-9349	TP Electric (Lawrenceville IL)
800-358-3927	Texas Electric Motors (Dallas TX)
800-899-8681	Texas Oil Electric (Abilene TX)
800-854-3302	US Electrical Motors (Santa Fe Springs CA)
800-243-2700	US Electrical Motors (Orange CT)
800-323-5223	US Electrical Motors (Chicago IL)
800-851-7713	US Electrical Motors (St Louis MO)
800-683-4126	US Electrical Motors (Dallas TX)
800-762-0909	Universal Motor Distributors (Philadelphia PA)
800-231-4945	Warfield Electric (Ontario CA)
800-435-9351	Warfield Electric Co Inc (Frankfort IL)

800-435-9346	Warfield Electric Company Inc (Frankfort IL)
800-227-0835	Weeks Electric (Grand Rapids MI)
800-343-9996	Weg Electric Motors Corp (Boston MA)
800-451-8798	Westinghouse Motor Co (Round Rock TX)
800-247-6859	Westinghouse Motor Co (US)
800-648-4172	Whitfield Electric (Hanson KY)
800-398-7114	Wilde Electric Motor Supply (Hermiston OR)

Electrical Equipment & Supplies

800-634-7643	ABB Power Distribution (Florence SC)
800-422-2170	ABB Power T & D Parts & Service (Pittsburgh PA)
800-634-6005	ABB Power Transmission Inc (Allentown PA)
800-543-3395	ABC/ASI Inc (Saginaw MI)
800-328-7845	AG Electric Specialist (Racine MN)
800-428-4370	ARCO Electric Products Corp (Shelbyville IN)
800-367-8200	Abbott Technologies Inc (Sun Valley CA)
800-221-9574	Academy Electrical Products (Yonkers NY)
800-334-5214	Acme Transformer Division of Acme Electrical Corp (Lumberton NC)
800-874-2834	All American Semiconductor (Bedford MA)
800-523-9614	All Phase Electric Supply (Lehigh PA)
800-255-3828	Allwire (Fresno CA)
800-223-7487	American Bearing & Drives (Hayward CA)
800-535-6268	American Contractors (Ft Worth TX)
800-631-5310	American Electric (Cranbury NJ)
800-526-6492	American Electric (Pittsburgh PA)
800-221-5268	American Fitting (Westwood NJ)
800-532-7979	American Safety Utility Corp (Shelby NC)
800-421-2242	American Solenoid (Des Plains IL)
800-424-9473	Ancor Marine-Whsle Distributor (Cotati CA)
800-458-5578	Anderson Lyn (Orofino ID)
800-428-5050	Arc (Greenwood IN)
800-332-2201	Argo Systems (Sunnyvale CA)
800-643-8776	Arkansas Electric Coops (Little Rock AR)
800-242-6613	Arkansas Electric Coops (US)
800-482-1277	Arkansas Electric Coops (US)
800-552-1789	Arms Electric (Elkhart IN)
800-451-0869	Arnev Products (S Elgin IL)
800-367-8277	Arrow Electronics Inc (Tustin CA)
800-762-4750	Astro Industries Inc (Dayton OH)
800-462-3005	Atlantic Electric (Worcester MA)
800-438-2010	Atlantic Electric Systems (Charlotte NC)
800-331-2056	B & P Electrical Surplus (Baton Rouge LA)

800-451-2545	Bainbridge Electric Supply Inc (Bainbridge GA)
800-621-5541	Baron Wire & Cable (Skokie IL)
800-451-9753	Bellatrix Systems (Bend OR)
800-356-4266	Bender Inc (Exton PA)
800-527-1321	Bicks Supply (Fayetteville NC)
800-342-1115	Braid Electric (Nashville TN)
800-892-9246	Breaker & Control Inc (Houston TX)
800-527-5986	Breaker World (Bensenville IL)
800-392-6666	Brenner Electrical Sales Inc (Houston TX)
800-292-4358	Brownell Electro (Memphis TN)
800-222-7532	Butler Mfg Co (Parkersburg WV)
800-428-9267	CSE Technologies (New London MN)
800-542-0007	California Breakers (CA)
800-824-3197	California Tube Laboratory (Santa Cruz CA)
800-228-5123	Calterm Inc (El Cajon CA)
800-868-7459	Cameron & Barkley (Athens GA)
800-282-0435	Cameron & Barkley (Columbus GA)
800-826-9591	Cardello Electric Supply (Wheeling WV)
800-672-5874	Carvir (Goldsboro NC)
800-323-6696	Casey Bill Electric Sales (Bensenville IL)
800-562-0535	Cefco Division Of The English Electric Corp (Macon GA)
800-854-4294	Challenger Electrical Equip Corp (Orlando FL)
800-433-1043	Cinch Clamp (Reno NV)
800-241-7345	Cleveland Electrical Co (Atlanta GA)
800-282-7150	Cleveland Electrical Cor (US)
800-323-9355	Coleman Cable Systems Electrical Conductors Inc (N Chicago IL)
800-621-0987	Columbia Electric Supply (Aberdeen WA)
800-722-6638	Comet World (Hillside NJ)
800-292-8928	Condor Distributor (Blankeslee PA)
800-722-3726	Connersville Electric Supply (Connersville IN)
800-874-8746	Consolidated Electrical Distributors (Great Falls MT)
800-828-1169	Control Devices Inc (Flint MI)
800-257-0941	Coxco (Spokane WA)
800-742-4088	Curby Riss Supply Inc (Bloomington IN)
800-458-9600	D & F Liquidators (Hayward CA)
800-328-0298	DPM (Richfield MN)
800-543-7791	Daycoa (Medway OH)
800-342-3618	Demico Inc (Marietta GA)
800-222-2140	Desco Inc (Salisbury NC)
800-222-9148	Desco (Albemarle NC)
800-222-5845	Desco (Lexington NC)
800-332-9407	Diamond Electric Supply Co Inc (Gastonia NC)

800-835-7872	Dickman Supply (Sidney OH)
800-332-3827	Dimensional Sciences (Lakemont GA)
800-424-7791	Directelec (Little Canada MN)
800-272-2077	Divesco (Hot Springs AR)
800-521-5230	Dulmisons Electric Equip Transmission & Dist Power Lines (Lawrenceville GA)
800-347-0794	Dynamic Electric Supply Inc (Burlington NC)
800-835-8889	E & M Sales (Salt Lake City UT)
800-334-4900	E T C Equip Corp (Lynbrook NY)
800-824-1085	Eagle Engineering & Supply (Alpena MI)
800-247-0859	Eagle Instruments & Controls (Largo FL)
800-628-5788	East Coast Lamp Sales (E Northport NY)
800-441-1771	Eastco Intl (Chicago IL)
800-233-4876	Eaton (Lincoln IL)
800-323-6174	Electri-Flex Co (Roselle IL)
800-632-0268	Electric Supply & Equip (Greensboro NC)
800-332-3090	Electric Supply & Equip (Morrisville NC)
800-672-2173	Electric Supply & Equip (Rocky Mount NC)
800-362-6454	Electric Surplus (Ham Lake MN)
800-421-8855	Electric Switches (Los Angeles CA)
800-252-4640	Electric Switches (US)
800-231-2017	Electric Wire & Cable Co Inc (Houston TX)
800-328-0328	Electrical Advertiser (Brooklyn Center MN)
800-473-7661	Electrical Assemblies (Racine WI)
800-422-6077	Electrical Distributors (Atlanta GA)
800-638-8376	Electro Term Inc (Springfield MA)
800-253-3761	Electro-Mechanics Co (Austin TX)
800-343-2278	Electroline Manufacturer Co (Cleveland OH)
800-426-8247	Electronic Component Distributors (Victoria TX)
800-431-2308	Entrelec Inc (Irving TX)
800-233-7624	Epoch Distributors Inc (South Bend IN)
800-972-5109	Ericson Mfg (Willoughby OH)
800-476-2156	Eritech Inc (Aberdeen NC)
800-232-3726	Evansville Supply (Evansville IN)
800-231-4692	Excess Electrical Inc (Houston TX)
800-638-6594	F.I.C. Corp (Rockville MD)
800-251-9388	Fl Shock Inc Repair Center (Knoxville TX)
800-645-2230	Fairway Electrical RIC Corp (Amityville NY)
800-334-4940	Fayetteville Transformer Co (Fayetteville NC)
800-522-2281	Fife Custom Equip (Lakeland FL)
800-445-0916	Fistell Electronics Supply (Denver CO)
800-537-6887	Florida Heat & Control (Freeport FL)
800-552-0902	Foster Electric Co Inc (Lynchburg VA)

800-255-1919	Fuses Unlimited (Van Nuys CA)
800-833-2530	G & G Technology (Chandler AZ)
800-533-5885	GE Installations & Service Engineering Department (US)
800-243-7313	GE Supply (Indianapolis IN)
800-634-3726	GE Supply-East (Atlanta GA)
800-262-3114	GE Supply-West (Dallas TX)
800-325-0416	GNWC (Phoenix AZ)
800-424-3111	General Magnetic Electric Wholesale (CA)
800-356-2345	George (Fountain Valley CA)
800-752-0536	Glasco Electric (Joplin MO)
800-452-1676	Glaze Supply Co (Dalton GA)
800-453-1120	Gola Chestnut Inc (Philadelphia PA)
800-826-6318	Graybar (Phoenix AZ)
800-423-8540	H L C Electric Supply (Baldwin Park CA)
800-544-2105	HKL North America (Worthington OH)
800-327-9989	Hall Mark Electronics (Tulsa OK)
800-842-7437	Harger Lighting Protection Inc (Mundelein IL)
800-541-8910	Harold Electric Co (Walla Walla WA)
800-526-4182	Heyman Mfg (Kenilworth NJ)
800-527-0663	Hi-Line Electric Co (Dallas TX)
800-742-8054	High Voltage Test Lab (Wahpeton ND)
800-447-3150	Hillsdale Terminal & Tools (Jonesville MI)
800-346-4483	Hite Co Inc The (Olean NY)
800-423-4483	Hite The (Meadville PA)
800-367-1212	Holden H A (Minneapolis MN)
800-422-9132	Hood Electric (Youngstown OH)
800-247-8593	Hull Electric (Bradford PA)
800-231-0099	Indiana Coil Service (South Bend IN)
800-327-9548	Industrial Electric Sales & Service Inc (Hollywood FL)
800-647-8877	Innovative Technology (Brooksville FL)
800-922-2003	Interstate Electric Supply (Chantilly VA)
800-938-4343	J E Edwards Company Inc (Houston TX)
800-426-7021	Jenkins W L (Canton OH)
800-536-1757	Kay Electric (Altus OK)
800-466-6361	Kay Electric (Lawton OK)
800-521-8818	Kelley Allan (Barrington NH)
800-222-5129	Kent Industries (Sussex NJ)
800-322-5338	Kiefer Electrical Supply Co (Peoria IL)
800-642-3385	Kirby Risk Suply Co Inc (Anderson IN)
800-382-0886	Kirby Risk Supply Co Inc (Crawfordsville IN)
800-992-3740	Kirby Risk Supply (Bedford IN)
800-458-5255	Kirby Risk Supply (Kokomo IN)

800-832-5667	Kirby Risk Supply (Terre Haute IN)
800-262-0177	Kirby Risk Supply (Lima OH)
800-822-4749	Kulwin Electric Supply (Bloomington IN)
800-331-8172	Lakeland Engineering Equip (Denver CO)
800-423-6295	Lakeland Engineering Equip (US)
800-235-5253	Lakeland Engineering Equip (Omaha NE)
800-343-3256	Lee Electrical Enclosures (Everett MA)
800-233-6588	Lee Electrical Enclosures (US)
800-526-2703	Lenco Industrial Electrical Supplies (Warren MI)
800-453-8377	Levers W E & Associates (New Castle PA)
800-626-8145	Liberty Electronics Inc (Franklin PA)
800-827-3253	Light Circuit Breaker Inc (Miami FL)
800-526-2731	Lightolier Controls (Garland TX)
800-654-9928	Lone Star Electric (Odessa TX)
800-344-6180	Lowndes Electrical Supply (Valdosta GA)
800-447-3748	Loyal Supply (Albany NY)
800-858-4205	Lubbock Electric Co (Lubbock TX)
800-458-2569	Luckow Circuit Breaker (Santa Ana CA)
800-237-4965	Luckow Electric (Van Nuys CA)
800-526-2600	Lupia Electric Supply Inc (Ogdensburg NY)
800-462-3600	Lupia Electric Supply Inc (Watertown NY)
800-233-1666	M & A Electrical Parts Remanufacturer (Phoenix AZ)
800-541-9997	Magnetek Inc (Los Angeles CA)
800-624-6383	Magnetek Inc (La Vergne TN)
800-526-4343	Major Electrical Supplies (Belle Glade FL)
800-221-5722	Martel Electronics Corp (Windham NH)
800-356-2486	Maydwell & Hartzell (Butte MT)
800-331-6681	Maydwell & Hartzell Utility Equip Co (Phoenix AZ)
800-331-3259	McCormick & Associates (Alpharetta GA)
800-581-0866	McJunkin Corp (La Marque TX)
800-433-7642	Meltric Corp (Cudahy WI)
800-752-9312	Mercury Switches (Elkhart IN)
800-247-6005	Minnesota Electric Supply (Marshall MN)
800-992-8830	Minnesota Electric Supply (Willmar MN)
800-942-7561	National Circuit Breaker (Santa Fe Springs CA)
800-435-4481	Nehring Electrical Works (De Kalb IL)
800-431-6272	Nelson Electric Supply (Racine WI)
800-272-2603	New Entertainment Satellite Systems Inc (Pittsfield NH)
800-323-0588	Nichifu America (Elk Grove Village IL)
800-362-6331	Norstan Electronics Inc (Santa Ana CA)
800-647-0220	North Coast Electric (Longview WA)
800-533-6377	Number 1 Sales (Richfield OH)

800-446-4085	OEM Response Center Order Assistance (Pittsburgh PA)
800-621-4022	PM Sales (Chicago IL)
800-225-4022	PM Sales (US)
800-662-2290	Panel Components Corp (Santa Rosa CA)
800-255-0981	Panelboard Specialties Whsle Electric Inc (Lake Elsinore CA)
800-331-1345	Payne Engineering (Scott Depot WV)
800-526-4198	Peach State Instrument Co (Marietta GA)
800-582-0640	Penz Enterprises (Boynton Beach FL)
800-342-6035	Petrick Expediting UEC (Verona PA)
800-323-9562	Phoenix (Wood Dale IL)
800-421-6511	Pico Macom Inc (Lake View Terrace CA)
800-428-0431	Platt Electric Supply (Rancho Cordova CA)
800-643-2559	Platt Electric Supply (Sacramento CA)
800-257-5288	Platt Electric Supply (Beaverton OR)
800-992-3070	Platt Electric Supply (Seattle WA)
800-548-6728	Positive Technologies Inc (San Antonio FL)
800-341-5327	Power Line Supply (Lima OH)
800-237-7653	Power Line Supply (US)
800-451-1900	Precision Valley Electrical (N Springfield VT)
800-532-0955	Presco Electric (Flagstaff AZ)
800-992-1555	Progress Electric Supply (Oak Park MI)
800-645-7902	R & S Maintenance (Taylor MI)
800-874-8037	RF Prototype Systems (San Diego CA)
800-252-3411	RSI (Pflugerville TX)
800-522-9108	Rafalian Ben (Culver City CA)
800-334-6475	Ranger Communications (San Diego CA)
800-332-1111	Rapid Power Technologies Transformer Division (Brookfield CT)
800-462-5560	Rawlinson Electric Supply (Dallas TX)
800-632-7237	Reed City Power Line Supply (Reed City MI)
800-327-6445	Relay & Control (Hauppauge NY)
800-526-5376	Relay Specialties (Oakland NJ)
800-443-8084	Richards Supply (Chillicothe OH)
800-282-4593	Richards Supply (New Boston OH)
800-848-4583	Richards Supply (US)
800-521-7829	Rinky Dinks (New Albany IN)
800-637-4425	Rittal (Springfield OH)
800-553-8690	Robinson B A Electrical Co (Lincolnton NC)
800-824-9450	Rosemount Inc (Baton Rouge LA)
800-654-3205	Ross Engineering Corp (Campbell CA)
800-843-2764	Schweber Electronics (Englewood CO)
800-634-5638	Scott Parish Electrical Supply Of Raleigh North Carolina (Raleigh NC)
800-258-5559	Seamans Supply (Manchester NH)

800-562-1160	Seamans Supply (Manchester NH)
800-225-4864	Seimens Distribution Equip (La Mirada CA)
800-547-2281	Selectric (Umatilla OR)
800-435-4066	Siemens/New Lenox-Distributor (Joliet IL)
800-749-2825	Sights & Sounds (Alva OK)
800-323-5744	Simpson Electric (Elgin IL)
800-346-4927	Small Precision Tools & Sales (Petaluma CA)
800-645-0830	Snyder Electric (Lancaster OH)
800-438-8364	Southeastern Electric Range Parts Mfg (Gastonia NC)
800-633-8796	Southern Control Inc (Montgomery AL)
800-392-5770	Southern Control Inc (US)
800-633-4620	Southern Control Inc (US)
800-362-2788	Southwest Electric (Memphis TN)
800-255-9277	Southwestern Electric Wire & Cable (Irving TX)
800-772-0138	Southwestern Electric Wire & Cable (Irving TX)
800-423-2440	Splane Electric Supply Company (Van Buren MI)
800-922-5598	Spotter Controls (Rockwall TX)
800-634-2003	Square D Technical Services (Schaumburg IL)
800-232-9410	Standard Electric Co (Bad Axe MI)
800-992-0005	Starbuck Sprague Co The (Waterbury CT)
800-258-7415	State Electric Supply (Ashland KY)
800-443-9044	State Electric Supply (Pikeville KY)
800-544-6755	State Electric Supply (US)
800-874-7856	State Electric Supply (US)
800-642-8642	State Electric Supply (Bluefield WV)
800-624-3417	State Electric Supply (Huntington WV)
800-642-2762	State Electric Supply (Huntington WV)
800-642-2630	State Electric Supply (Parkersburg WV)
800-626-1223	Steel Electric Products Co Inc (Brooklyn NY)
800-621-3335	Stranco Products (Elk Grove Village IL)
800-626-0497	Stusser (Seattle WA)
800-545-7778	Superior Lighting (Ft Lauderdale FL)
800-445-8834	Supro Lux Mfg (La Jolla CA)
800-528-9541	Sweetco Sales (Claremore OK)
800-325-2443	Swofford Electric Supplies Corp (Grand Prairie TX)
800-553-2425	T-Electra (Euless TX)
800-452-8409	Taylor Electric Supply (Portland OR)
800-547-0922	Taylor Electric Supply (Portland OR)
800-821-9066	Telecom Electric Supply (Plano TX)
800-782-9096	Telecom Electric Supply-Fax (Plano TX)
800-332-9149	Tennessee Valley Electric Supply (Knoxville TN)
800-626-3667	Tico Products (Fraser MI)

800-642-7467	Tolley Electric (Clarksburg WV)
800-223-6564	Top In Sound (Anderson IN)
800-448-5636	Torq Electric Equip (Detroit MI)
800-662-3422	Tower Electronics (Green Bay WI)
800-345-2279	Townsend Electric Inc (Townsend MT)
800-824-8282	Trans Coil (Milwaukee WI)
800-646-6755	Trans-World Electric (Port Arthur TX)
800-327-6903	Transpo Products (Orlando FL)
800-428-4449	Transportation Safety Devices Inc (Indianapolis IN)
800-245-4552	Transtech Of South Carolina (Piedmont SC)
800-231-5082	Tristate Electrical Supply (Baltimore MD)
800-458-9862	Tristate Electrical Supply (York PA)
800-282-7985	Tristate Utility Supply Products (Marietta GA)
800-255-6630	US Electric Supply (Atlanta GA)
800-245-6378	US Safety Trolley (Bridgeville PA)
800-422-4260	Unicorn Electrical Products (Anaheim CA)
800-527-1279	Unitron (Dallas TX)
800-321-2192	Valen Mfg (Cleveland OH)
800-772-2942	Van Meter (Waterloo IA)
800-826-6839	Vantex Electric Products (Ft Worth TX)
800-457-8725	Vogel Enterprises - Telescopes Inc (Batavia IL)
800-468-3538	WESCO Data Telecommunications Products (Wilmington DE)
800-334-0550	Ward Transformer (Raleigh NC)
800-334-9600	Ward Transformer (US)
800-257-0038	Weaver Electric (Denver CO)
800-221-3595	Wemco (St Clair Shores MI)
800-522-6880	Wesco (Sterling VA)
800-344-0113	Western Enterprises (CA)
800-547-5101	Western Power Products Inc (Hood River OR)
800-547-9604	Western States Electric (Spokane WA)
800-525-6821	Westinghouse Network Protectors (Greenwood SC)
800-843-9695	Westnet Systems (Santa Ana CA)
800-342-3483	Wetherbee Electric Supply Co (Albany GA)
800-257-8182	Wheatland Tube Co (Collingswood NJ)
800-453-9473	Wrightco Wiring Products (Rancho Cordova CA)
800-841-3278	Wyrecart (Raleigh NC)
800-632-9090	Yesco (York PA)
800-882-8020	Zierick (Mt Kisco NY)

Electronic Equipment & Supplies - Wholesale & Manufacturers

800-468-6275	Alignrite (Burbank CA)
800-279-6835	Asymtek (Carlsbad CA)

800-831-7621	Austin Town Electronic Whsle (Youngstown OH)
800-799-9225	C.B. Shop (Florence CO)
800-222-2717	Carlin Systems (Ronkonkoma NY)
800-529-9483	Cole's Consultants (Mililani Town HI)
800-548-1657	Communicore (Sanford FL)
800-860-8910	Dancraft Enterprises (Inglewood CA)
800-649-2097	Delmo Inc (Fisk MO)
800-787-0575	E B W Electronics (Holland MI)
800-553-5228	Electrocatalytic (Warren NJ)
800-995-5076	Entec Group (Syracuse NY)
800-539-5377	Essex Electronics Inc (Carpinteria CA)
800-422-2801	Fourakre Electronics Inc (Valdosta GA)
800-424-0746	Fourakre Electronics Inc-Fax (Valdosta GA)
800-727-2297	Fuller's Whsle Electronics (US)
800-666-2296	Fuller's Whsle Electronics (Parkersburg WV)
800-640-9948	Futura Components (College Point NY)
800-682-2778	Giltronics (Kapaa HI)
800-486-0726	Holmes Distributors (Portland OR)
800-941-7123	Horseshoe Wire & Cable (Altoona PA)
800-926-4726	Instrument Research (Columbia MD)
800-345-1365	Integrated Network Corp (Bridgewater NJ)
800-662-5515	Integrated Network Corp (Bridgewater NJ)
800-336-5226	Laco Electronics (Decatur IL)
800-409-3009	Manistar Electronics (Laguna Hills CA)
800-577-7576	Mohawk Wire & Cable Corp (Schaumburg IL)
800-831-9172	Murata Electronics Of North America Inc (GA)
800-808-4686	Newtech Electronics (Hauppauge NY)
800-999-4998	Patriot Circuit Breaker & Controls (Bedford NH)
800-644-7501	Pioneer Technology Group (Bellevue WA)
800-347-0900	Powell Electronics (Philadelphia PA)
800-219-8900	Process Electronics (Oneida WI)
800-732-3457	Rag Electronics Inc (Newbury Park CA)
800-673-5700	Relay Specialties (Haddon Heights NJ)
800-222-2810	Reliable Communications Inc (Angels Camp CA)
800-283-2636	Scientific Atlanta (Melbourne FL)
800-247-9106	Service Electronics Fax Line (Eden Prairie MN)
800-664-0495	Sunward Electronics Inc (Delmar NY)
800-441-1139	Symbios Logic Inc (Stoneham MA)
800-543-1727	Syntrex (Wall Township NJ)
800-352-3493	Technomark (Minneapolis MN)
800-240-3788	Telexport (Hialeah FL)
800-440-3211	Tescon America Inc (Fremont CA)

800-839-6283	Texmate (Vista CA)
800-879-4963	Toshiba America Electronic Components (Irvine CA)
800-292-7736	Warren Processing Lab (Brooklyn NY)
800-621-0049	Wiremold (W Hartford CT)
800-947-3665	Wiremold (US)

Electronic Testing Equipment

800-531-1148	A P I (Mountain View CA)
800-531-1149	A P I-Fax Only (San Jose CA)
800-532-3381	AT&T Capital Corporation (US)
800-524-0747	Accutest Instruments Inc (NJ)
800-404-2832	Advanced Test Equipment Corp (US)
800-251-0706	Atlantic Coast Instruments (Wall NJ)
800-752-8272	Baker Instrument Co (Ft Collins CO)
800-346-9906	Certitech Corp (Irvine CA)
800-832-0732	Datron Technology (Farmington MI)
800-854-1566	Digimax Instrument (San Diego CA)
800-528-7405	Direct Safety Company (AZ)
800-372-6832	Dranetz Technologies (Edison NJ)
800-254-9478	EMC Automation (Austin TX)
800-729-8084	EQS Systems (Chesterland OH)
800-262-9606	Electra Sound (Cleveland OH)
800-421-2968	Electronic Devices Inc (Chesapeake VA)
800-348-7242	Electronic Instruments USA Inc (Stamford CT)
800-957-3342	Fluke/Zack Electronics (US)
800-359-8570	Frontline Test Equip (Oak Brook IL)
800-437-3687	GE Capital Computer Rental Services (US)
800-443-6723	Genrad Inc (Concord MA)
800-348-3596	Get Control (Tempe AZ)
800-343-1344	Graham-Lee Electronics (Excelsior MN)
800-327-9308	Helper Instruments (Indian Harbour Beach FL)
800-445-3835	Hilevel Technology (Tustin CA)
800-566-1818	I T C Instruments Technology (Carson City NV)
800-952-4499	ICS Electronics (Milpitas CA)
800-654-7368	Industrial Resources (Brea CA)
800-345-6140	Instrument Repair Labs (Broomfield CO)
800-826-3759	Instrument Repair Service (Norcross GA)
800-635-3030	JW Harley Inc (Twinsburg OH)
800-552-1115	Keithley Instruments Inc (Cleveland OH)
800-645-5104	Leader Instruments Corp (Hauppauge NY)
800-237-5964	Magni Systems (Beaverton OR)
800-821-0023	Martel Electronics Corp (Salem NH)

800-289-6734	Max Sands Enterprises Inc (N Hollywood CA)
800-242-8051	Metalink Corp (Chandler AZ)
800-800-5003	Metermaster (Addison IL)
800-432-3424	Metric Equipment Sale (Forster City CA)
800-654-5659	Micromanipulator The (Carson City NV)
800-992-9943	Mouser Electronics (Mansfield TX)
800-336-7723	Naptech (Cobb CA)
800-593-7617	National Calibration & Testing Laboratories (Minneapolis MN)
800-262-2296	Pan American Systems Corp (Virginia Beach VA)
800-468-4120	Power Monitors Inc (Harrisonburg VA)
800-772-1519	Probe Master Inc (El Cajon CA)
800-835-1099	Q Corp (Derby KS)
800-391-3457	Rag Electronics Inc (Newbury Park CA)
800-635-9591	Red Ball Electronic Repair (Pembroke Pines FL)
800-233-5807	RenTelco (Richardson TX)
800-548-6305	Rod L Electronics (Menlo Park CA)
800-880-0048	Schaffner EMC-West Coast Office (Irvine CA)
800-527-8894	Scherrer Instruments (St Louis MO)
800-922-2969	Sotcher Measurement Inc (Sunnyvale CA)
800-443-4132	Sparton Electronics (Brooksville FL)
800-645-5398	Sperry A W Instruments (Hauppauge NY)
800-344-8131	Tau-Tron Inc (Westford MA)
800-451-7368	Technilease Corp (Newark NJ)
800-995-4665	Tektest (Arcadia CA)
800-847-0407	Tektronix Inc (Houston TX)
800-426-2200	Tektronix (Beaverton OR)
800-835-6100	Tektronix Inc (US)
800-835-6244	Tektronix Inc (US)
800-957-3340	Tektronix/Zack Electronics (US)
800-835-2606	Telogy (Redwood City CA)
800-558-5080	Temptronic Corporation (Newton MA)
800-227-1995	Test Equip Co (Mountain View CA)
800-615-8378	Test Equip Connection Corp (Hampton Base NY)
800-832-3637	Test Equipment Corporation (US)
800-442-5835	Test Lab Co (Mountain View CA)
800-368-5719	Test Probes Inc (San Diego CA)
800-527-4642	Tucker Electronics (Garland TX)
800-228-8574	VLSI Standards Inc (Roseveille CA)
800-548-9806	Valhalla Scientific (San Diego CA)
800-321-6384	Vitec, Inc (Cleveland OH)
800-622-5515	Wavetek Communication-CATV Sales (Indianapolis IN)
800-851-1202	Wavetek Communications-Componet Sales & Mktg (Indianapolis IN)

800-851-1198	Wavetek Communications-Customer Service (Indianapolis IN)
800-245-6356	Wavetek Communications-Mobile Sales (Indianapolis IN)
800-547-4515	Westcon Inc (Portland OR)
800-423-0268	Western Industrial Service (Inglewood CA)

Fiber Optics

800-998-3947	AT&T Authorized Stocking Distributor/Zack Electronics (US)
800-353-4413	Atlanta Cable Sales Inc (Norcross GA)
800-545-4306	BT & D Technologies (San Jose CA)
800-274-8397	Brite Expectations (Casselberry FL)
800-833-4237	Dolan Jenner Industries (Lawrence MA)
800-500-0347	Fiber Instrument Sales Inc (NY)
800-894-9694	Fiberdyne Labs (Herkimer NY)
800-50FIBER	Fiber Optic Association (Boston MA)
800-874-8657	Fiberoptics Specialties Inc (Palmetto FL)
800-543-2558	Fiberoptics Technology (Pomfret CT)
800-637-8254	Fotec (Medford MA)
800-642-7841	General Connector (San Pedro CA)
800-962-5327	George Ingraham Corporate Offices & Accounting (Stone Mountain GA)
800-544-6853	Gould Fiber Optics Operations (Millerville MD)
800-848-4527	Heraeus Amersil (Duluth GA)
800-821-6363	Hoechst Celanse Fiber Optic Products (Summit NJ)
800-622-6702	Indian Nations Fiberoptics (Sulphur OK)
800-553-5554	Lazarus Lighting Design (Glendale CA)
800-871-9838	Lightwave Communication (Milford CT)
800-445-4617	Litra Inc (Norcross GA)
800-456-3766	Meson Design And Development (Binghamton NY)
800-342-5256	National Communications Inc (Newark NJ)
800-228-4692	Northern Wire & Cable Inc (Troy MI)
800-622-7711	Optical Cable Corporation (VA)
800-523-8917	Osburne Associates (Logan OH)
800-433-5711	Petroflex NA Inc (Gainesville TX)
800-314-1104	Porta Systems Corp (Fayetteville GA)
800-589-2556	Protech (Ft Worth TX)
800-743-6271	Rifocs (Camarillo CA)
800-321-1690	Sales Associates Picture Frames & Mouldings (Cleveland OH)
800-225-7977	Schlafli Engineering (Palmyra NJ)
800-322-3116	Siecor/Emergency Fiber Optic Cable Service (Hickory NC)
800-438-2627	Siecor/Hardware & Test Equip (Hickory NC)
800-245-4225	Siecor/IBM System Cabling Products (Hickory NC)
800-358-7378	Sumitomo Electric Fiber Optics Corp (Research Triangle Park NC)
800-447-3213	Telecom Engineering Consultants (Miami FL)

Lighting Fixtures - Wholesale & Manufacturers

800-373-1549	Alert Lite Neon (Sun Valley CA)
800-577-5483	Amalgamated Electrical Brokers LTD (Denver CO)
800-553-7722	Cornelius Architectural Products (Pittsburgh PA)
800-446-3262	Danalite (Compton CA)
800-843-1602	Design Lighting Products (Mesa AZ)
800-522-2626	Elco Lighting (Culver City CA)
800-421-8226	Elsco Lighting (US)
800-343-6253	Emerald Bird Caddy (Eugene OR)
800-809-9063	Galaxy Electrical (Hialeah FL)
800-966-4375	Gaspen Vitus & Associates (Phoenix AZ)
800-783-1846	Guyco (Miami FL)
800-437-8498	Inlite Corp East (Hight Point NC)
800-527-0998	Lavery Lighting (Corona CA)
800-524-5267	Light Savers (New York NY)
800-842-6113	Lighting Plastics (Van Buren AR)
800-932-0436	Lightolier Corp (Fall River MA)
800-215-1068	Lightolier (Secausus NJ)
800-648-4141	Michael Lynn Lighting Galleries (Flint MI)
800-234-3724	Ott Light Systems Inc (Santa Barbara CA)
800-553-6167	Power Lighting (Mt Vernon NY)
800-783-9920	Secure Lighting (Miami FL)
800-544-4877	Superior Lamp & Electrical Supply Co Inc (New York NY)
800-422-6000	USA Electric (Los Angeles CA)
800-426-9977	USA Electric (Los Angeles CA)
800-334-1265	Westek Associates (San Diego CA)

Wire & Cable - Electric

800-832-3340	2nd Source Wire & Cable (Santa Fe Springs CA)
800-451-7292	A Z Industries Inc (Northbrook IL)
800-432-0325	Accu-Tech Corp (Austin TX)
800-227-0628	Accu-Tech Corp (US)
800-493-8328	Accu-Tech Corporation (Tampa FL)
800-221-4767	Accu-Tech Corporation (Atlanta GA)
800-835-4949	Adelphia Cable (Blairsville PA)
800-237-4542	Adirondack Wire & Cable (Woonsocket RI)
800-626-3608	Advanced Computer Cables Inc (Hialeah FL)
800-257-8167	Aetna Insulated Wire (Bellmawr NJ)
800-433-9903	Alcon Corp (Los Angeles CA)
800-443-6481	Alconex Specialty Products (Ft Wayne IN)
800-621-5837	All States Inc (Chicago IL)
800-522-5742	Alpha Wire (Elizabeth NJ)

800-225-8588	American Flexible Conduit (New Bedford MA)
800-521-9903	Anicom (Tampa FL)
800-231-4322	Anicom (Tucker GA)
800-252-9895	Anicom (Mt Prospect IL)
800-543-5810	Astro Industries Inc (Dayton OH)
800-752-5601	Astro Industries New England Division (Pawtucket RI)
800-353-4413	Atlanta Cable Sales Inc (Norcross GA)
800-423-4659	Atlas Wire & Cable (Montebello CA)
800-552-0661	Bay Associates Electrical Wire And Cable (Menlo Park CA)
800-468-7309	Bay State Wire & Cable (Lowell MA)
800-233-3539	Bee Wire & Cable (Ontario CA)
800-998-3948	Belden Electronic Wire & Cabel/Zack Electronics (US)
800-235-3361	Belden (Naperville IL)
800-235-3364	Belden Wire & Cable (Richmond IL)
800-581-9873	Better Source The (Chicago IL)
800-645-3505	Birnbach Co (Farmingdale NY)
800-423-2322	CZ Labs (Congers NY)
800-346-6211	Cable & Harness Fabricators (Windsor VT)
800-422-2539	Cable Exchange (Garden Grove CA)
800-328-8116	Cable Master (Menomonee Falls WI)
800-292-9473	Cablec Continental Cables (York PA)
800-227-9473	Capital Electric Wire & Cable (Waukesha WI)
800-282-8188	Cardinal Supply Inc (Warminster PA)
800-227-5953	Carlyle Inc (Tukwila WA)
800-472-2054	Carpenter Cable (London KY)
800-262-5999	Cat Wire & Cable Corp (Miami FL)
800-354-9473	Coastal Wire & Cable (Houston TX)
800-323-1403	Cole Wire & Cable Co (Lincolnwood IL)
800-228-1407	Commercial & Industrial Electrical Sales Inc (Pelham AL)
800-222-4597	Crown Wire & Cable Co Inc (Millville NJ)
800-832-3600	Custom Cable Corporation (Westbury NY)
800-446-2232	Custom Cable Industries (Tampa FL)
800-323-0198	Dearborn Wire & Cable L.P. (Wheeling IL)
800-448-1711	Del City Wire Co Inc (Oklahoma City OK)
800-451-1376	Diversified Wire & Cable (Troy MI)
800-440-0409	Durable Products (San Jose CA)
800-962-3588	Elecon Wire & Cable (Medley FL)
800-962-9473	Encore Wire (McKinney TX)
800-826-1265	Energy Electric Assembly Inc (Auburn Hills MI)
800-332-2412	Energy Electric Cable (Pensacola FL)
800-325-2245	Energy Electric Cable Inc (Los Angeles CA)
800-233-9959	Energy Electric Cable Inc (Ft Lauderdale FL)

800-532-2353	Energy Electric Cable Inc (Orlando FL)
800-972-9049	Energy Electric Cable Inc (Chicago IL)
800-521-6521	Energy Electric Cable Inc (Auburn Hills MI)
800-521-6520	Energy Electric Cable Inc (Sterling Heights MI)
800-368-8874	Essex Group Inc (IN)
800-227-8001	FJC Electronics (Clinton MA)
800-245-9473	Fay Electric Wire (Elmhurst IL)
800-472-3780	Fay Electric Wire (Oshkosh WI)
800-642-8810	Ferreira Industries (Tracy CA)
800-645-9473	First Capitol Wire & Cable (York PA)
800-322-3660	Ford Wire & Cable Corp (Vero Beach FL)
800-451-3788	Formulabs Industrial Inks (Piqua OH)
800-772-5951	Fort Worth Wire & Cable (Ft Worth TX)
800-447-4071	Futronix (Houston TX)
800-232-4692	GNWC Inc (Downers Grove IL)
800-468-2121	Great Northern Wire And Cable (Downers Grove IL)
800-445-6905	Group L Communications Inc (Nashville TN)
800-822-8919	Hall Wire & Cable (Cedarhurst NY)
800-225-6933	Houston Wire & Cable (Houston TX)
800-468-9473	Houston Wire & Cable (TX)
800-262-8947	IMS Inc (Norwalk CT)
800-446-5612	IPC Cable (Houston TX)
800-762-7668	Ideal Wire Products Inc (Tewksbury MA)
800-523-3930	Imperial Wire & Cable Co Inc (Newtown PA)
800-800-1003	Insulectro (Lake Forest CA)
800-527-0010	Interstate Wire Co Inc (Dallas TX)
800-442-0073	Interstate Wire (Dallas TX)
800-819-2042	J I T Supply (Las Vegas NV)
800-633-3655	JT & T Proucts Corp (San Jose CA)
800-222-2254	Janor Wire & Cable (Ivyland PA)
800-523-2467	Janor Wire & Cable (US)
800-633-6339	K L Tannehill Inc (Minneapolis MN)
800-346-8734	Keystone Cable (Philadelphia PA)
800-528-0449	King Wire & Cable Corp (Phoenix AZ)
800-752-9233	Komar Mfg (Claysburg PA)
800-447-0124	L F Gaubert & Co (Peoria IL)
800-831-7534	L F Gaubert And Co Inc (New Orleans LA)
800-822-8808	Long Island Wire & Cable (Islip NY)
800-423-5097	MWS Wire Industries (Westlake CA)
800-392-4928	Manhattan Electric Cable Corp (Houston TX)
800-238-9473	McCormick And Assoc (Salisbury NC)
800-633-1432	Metro Wire & Cable (Norcross GA)

800-524-7444	Miles Tek (Argyle TX)
800-258-6922	Minnesota Wire & Cable Co (St Paul MN)
800-422-9961	Mohawk Wire & Cable Corp (MA)
800-346-6626	Montrose Products Co (Auburn MA)
800-227-7240	National Communications Inc (Newark NJ)
800-726-8044	National Supply (Green Bay WI)
800-645-1715	Nek Cable Inc (Bohemia NY)
800-522-2253	Nemal Electronics Inc (N Miami FL)
800-752-5855	Netcom Plus Wire & Cable Inc (Rochester Hills MI)
800-243-3408	New England Insulated Wire (Berlin CT)
800-969-5106	Nichifu America (Elk Grove Village IL)
800-322-3785	Nicholson Sales & Paige Electric (Lubbock TX)
800-451-6852	Northeast Wire & Cable Inc (Niagara Falls NY)
800-228-4692	Northern Wire & Cable Inc (Troy MI)
800-860-7843	Northern Wire & Cable (Troy MI)
800-445-4623	Northern Wire & Cable(Cleveland OH)
800-445-4630	Northern Wire & Cable (US)
800-821-1028	Olympic Wire & Cable Corp (Charlotte NC)
800-526-2269	Olympic Wire & Cable Corp (Fairfield NJ)
800-331-9896	Olympic Wire & Cable (Ft Mill SC)
800-338-2536	Omega Leads (Santa Monica CA)
800-622-7711	Optical Cable Corporation (VA)
800-432-0459	Pacer Electronics (Ft Lauderdale FL)
800-527-3977	Pacer Electronics (Merritt Island FL)
800-634-5031	Pacer Electronics-Marine Dept (Ft Lauderdale FL)
800-892-3349	Page Mfg (Charlton MA)
800-742-5628	Phalco/CDT Cable Corp (Manchester CT)
800-421-3547	Philatron Intl MBE (Santa Fe Springs CA)
800-238-7514	Power & Telephone Supply (Memphis TN)
800-221-2305	Power Cords & Cables (College Point NY)
800-952-3842	Prestolite Wire Corp (Paragould AR)
800-589-2556	Protech (Ft Worth TX)
800-762-2253	Rapid Mfg (Orange CA)
800-331-2898	Red Wing Products Inc (Kellyville OK)
800-762-5666	Resource Wire & Cable Inc (Norcross GA)
800-729-2812	San Juan Assembly (Blanding UT)
800-642-7339	Saxton Wiring Cable (Sanford NC)
800-633-7210	Sea Wire & Cable (Madison AL)
800-231-9473	Service Wire Co (US)
800-624-3572	Service Wire Co (US)
800-831-4600	Service Wire Co (US)
800-624-3535	Service Wire Co (Huntington WV)

800-778-9473	Service Wire Sales (Pittsburgh PA)
800-792-9473	Southern Wire & Cable (Charlotte NC)
800-323-1334	Specialty Wire & Cable (Skokie IL)
800-348-3217	Stranco Products (Michigan City IN)
800-525-3851	Standflex Division M S W (Oriskany NY)
800-426-5367	Sunshine Wire & Connector (Miami FL)
800-932-9473	Sunshine Wire & Connector (US)
800-992-1073	T W Communication Corp (Cranford NJ)
800-553-1992	TPC Wire & Cable (Cleveland OH)
800-638-5505	TW Communications Corp (Orlando FL)
800-247-9099	Tappan Wire & Cable (Tappan NY)
800-821-3600	Taylor Cable Products (Grandview MO)
800-782-8068	Telacom Wire And Cable Co (Sultan WA)
800-331-6888	Tele Rep Communications (Port Jefferson Station NY)
800-458-9473	Telemarketing Inc (Auburn PA)
800-392-9842	Texas Electric Insulated Cable Corp (Houston TX)
800-872-8585	Texas Wirehouse (Houston TX)
800-825-9973	Texcan Cables (Pico Rivera CA)
800-765-7765	Texcan Cables (Sparks NV)
800-659-6362	Texcan Cables (Bellevue WA)
800-556-7770	Triangle Wire & Cable (Lincoln RI)
800-237-7528	Triangle Wire & Cable-Westwire Products (Glendale AZ)
800-367-2673	Unicable Inc (Cypress CA)
800-888-0738	Washington Cable Supply Inc (Lanham MD)
800-245-4964	West Penn Wire (Washington PA)
800-634-9473	West Penn Wire (Washington PA)
800-822-9473	Whitmore Wirenetics (Valencia CA)
800-323-0479	Whitney Blake Co Of Vermont (N Walpole NH)
800-433-9473	Wireman Inc The (Landrum SC)
800-325-9473	Wirexpress (Elk Grove Village IL)
800-932-9474	Wolverine Electronics (Warren MI)
800-851-9644	Zip Tape Identification Systems (Tempe AZ)

CHAPTER 13
GLOSSARY

10Base2 - 10 Mbps, baseband, in 185 meter segments. The IEEE 802.3 substandard for ThinWire, coaxial, Ethernet.

10BaseT - 10 Mbps, baseband, over twisted pair. The IEEE 802.3 substandard for unshielded twisted pair Ethernet.

110 Type Block - A wire connecting block that terminates 100 to 300 pairs of wire. It has excellent electrical characteristics and relatively small foot print. It organizes pairs horizontally.

66-Type Block - A type of wire connecting block that is used for twisted pair cabling cross connections. It holds 25 pairs in one to four vertical columns.

80x86 - Family of microprocessors made by Intel used in PC and clone computers.

A/D converter - Analog to digital converter.

Absorption - Loss of power in an optical fiber, resulting from conversion of optical power into heat and caused principally by impurities, such as transition metals and hydroxyl ions, and also by exposure to nuclear radiation.

Acceptance angle - the half-angle of the cone within which incident light is totally internally reflected by the fiber core. It is equal to arcsin (NA).

Acceptance Test - A test run on a host or network to determine its operation is satisfactory prior to acceptance by the purchaser or the purchaser's agent.

Access Charge - A charge, set by the FCC (Federal Communications Commission), for access to a local carrier by a user or long- distance supplier.

Access Code - Digits a user must enter to obtain access to a particular service, physical area or system.

Access Control Byte - Token ring field that holds the priority and reservation bits for a token or data packet.

Access (noun) - Level of authority or point of presence a user has for reading data.

Access (verb) - To connect, as in: to data in memory; to a peripheral (or another device) for use; to another host.

Access floor - A system of raised flooring that has completely removable and interchangeable floor panels. The floor panels are supported on adjustable pedestals or stringers (or both) to allow access to the area beneath.

Access Line - A line or circuit that connects a customer site to a network switching center or local exchange. Also known as the local loop.

Access Method - (1) A routine that prepares data for transmission. (2) A way of connecting to a host or peripheral. Also known as access routine.

Access Time - The period between a request to store or read data and the completion of that storage or retrieval.

Accessible - Can be removed or exposed without damaging the building structure. Not permanently closed in by the structure or finish of the building.

Accessible, readily (readily accessible) - Can be reached quickly, without climbing over obstacles or using ladders.

Active Device - A device with its own power source.

Active Hub - A device used to amplify transmission signals in certain network topologies. An active hub can be used to either add additional workstations to a network or to lengthen the cable distance between nodes (workstations and/or file servers) on a network.

Adapter - A connectivity device that links different parts of one or more subsystems, systems, hosts or networks. For example: a network interface card may also be called a network adapter.

Add-on-board - An optional circuit board that conveniently modifies or enhances a personal computer's capabilities. See also Memory Board, Network Interface Board, Network Interface Card.

Address (computer) - The designation of a particular word location in a computer's memory or other data register.

Address - An identified name for a logical or physical item that can be used to connect to that item.

ADP - Automated Data Processing, Automatic Data Processing, Administrative Data Processing, Advanced Data Processing.

Aerial Cable - Telecommunications cable installed on aerial supporting structures such as poles, sides of buildings, and other structures.

Aggregate - A masonry substance that is poured into place, then sets and hardens, as concrete.

AI - Artificial Intelligence.

AIA - American Institute of Architects.

Air Handling Plenum - A designated area, closed or open, used for environmental air circulation.

Algorithm - A well-defined set of rules that solve a problem in a finite number of steps. For example, a full arithmetic procedure for determining retransmission wait time.

Alphanumeric - A character set that contains letters, numerals (digits), and other characters such as punctuation. Also used to identify one of such characters.

Alpha Test - The first test of newly developed software or hardware (usually in a laboratory). See also beta test.

Alternate Route - A secondary communications path used to reach a destination if the primary path is unavailable.

Alternate Routing - A way of completing connections that use another path when the previous circuit is unavailable, busy or out of service.

Alternate Use - The ability to switch communications facilities from one type to another, i.e., voice to data, etc.

Alternating current (AC) - Electrical current which reverses direction repeatedly and rapidly. The change in current is due to a change in voltage which occurs at the same frequency.

AM - Amplitude modulation.

Ambient temperature - The temperature of the surroundings.

American National Standards Institute (ANSI) - A private organization that coordinates some US standards setting. It also approves some US standards that are often called ANSI standards. ANSI also represents the United States to the International Standards Organization. See also: International Standards Organization

American Standard Code for Information Interchange (ASCII) - A standard character set that (typically) assigns a 7-bit sequence to each letter, number, and selected control character. Erroneously used now to refer to (8-bit) Extended ASCII. The other major encoding standard is EBCDIC.

American Wire Gauge (AWG) - A standard used to describe the size of a wire. The large the AWG number, the smaller (thinner) the described wire.

Ampacity - The amount of current (measured in amperes) that a conductor can carry without overheating.

Ampere (or amp) - Unit of current measurement. The amount of current that will flow through a one ohm resistor when one volt is applied.

Ampere-hour - The quantity of electricity equal to the flow of a current of one ampere for one hour.

Amplifier - A device that increases the voltage output (strength) of all signals received on a line. Amplifiers work on analog lines. Contrast with repeater.

Amplitude modulation - A transmission technique in which the amplitude of the carrier is varied in accordance with the signal.

Amplitude - The size, in voltage, of signals in a data transmission. This voltage level is known as the amplitude.

Amplitude distortion - An unwanted change in the signal voltage which alters the data transmitted.

Analog - A format that uses continuous physical variables such as voltage amplitude or frequency variation to represent information. Contrast with digital.

Angle of incidence - The angle that a light ray striking a surface makes with a line perpendicular to the reflecting surface.

Angular misalignment - The loss of optical power caused by deviation from optimum alignment of fiber to fiber or fiber to waveguide.

Annunciator - A sound generating device that intercepts and speaks the condition of circuits or circuits' operations. A signaling device that gives a visual or audible signal (or both) when energized.

Anode - The positive electrode in an electrochemical cell (battery) toward which current flows.

ANSI - See: American National Standards Institute.

ANSI Character Set - The characters that include the 128 character ASCII character set and the 128 character Extended ASCII character set.

ANSI-568 - Commercial Building Telecommunications Wiring Standard. See EIA-568

Append - To change a file or program by adding to its end.

Application Layer - The top-most layer (Layer-7) in the OSI Reference Model providing such communication services as electronic mail and file transfer. It is generally defined as the top layer of a network protocol stack.

Application - A program that performs a user function. Synonymous with program.

Approved - Acceptable to the authority that has jurisdiction.

Approved Ground - A grounding bus or strap in a building that is suitable for connecting to data communication equipment. It includes a grounding subsystem, the building's electrical service conduit and a grounding conductor. See also EIA 607 and the National Electrical Code.

Arbitration - A set of rules for determining handling and priority of multiple communication sessions. Arbitration is less desirable than negotiation.

Architecture (computer) - The conceptual design of computer hardware.

Architecture - The relationship of the physical parts of a computer or network that is typically labeled by that relationship. For example, the Motorola 68000 architecture.

Archive - A procedure for transferring data from a storage disk, diskette or memory area to an external or removable storage medium.

Armoring - An additional protection layer on a cable to provide increased protection for abnormal environments. Often made of plastic coated steel it may also be corrugated for flexibility.

Array - A collection of photovoltaic (PV) modules, electrically wired together and mechanically installed in their working environment.

Artificial Intelligence (AI) - A computer's ability to perform functions usually associated with human thought processes. For example, reasoning, learning, and correction or self-improvement. The current useable AI is Expert Systems.

ASCII - See American Standard Code for Information Interchange.

ASIC - Application Specific Integrated Circuit, A chip that is custom designed for a particular application.

Assembler - A language that uses symbolic machine language statements that have a one-to-one correlation to computer instructions.

Async - Asynchronous. A data transmission method that sends one character at a time. Contrasted with the synchronous methods, which send a packet of data and then resynchronize their clocks. Asynchronous also refers to commands, such as in windowing environment, that may be sent without waiting for a response from the previous command.

Asynchronous Transfer Mode (ATM) - A method for the dynamic allocation of bandwidth using a fixed-size packet (called a cell). ATM is also known as "fast packet" and is an emerging WAN and LAN standard.

AT Command Set - The de facto modem command set developed by Hayes Microcomputer Products, Inc.

ATM - Automated Teller Machine or see Asynchronous Transfer Mode.

Attenuation coefficient - Characteristic of the attenuation of an optical fiber per unit length, in dB/km.

Attenuation - A general term used to denote the loss in strength of power between that transmitted and that received. This loss occurs through equipment, lines,or other transmission devices. It is usually expressed as a ratio in dB (decibel).

Attenuation Characteristic - As a signal travels on a cable, it gets weaker or attenuates. The attenuation characteristic of the medium is the rate at which it gets weaker.

Attenuator - A device that reduces signal power in a fiber optic link by inducing loss.

Authenticate (verb) - The function of verifying the identity of a person or process.

Authentication - (1) The verification of the identity of a person or process. (2) the code used to identify a person or process.

Authorization - Determining if a person or process is able to perform a particular action. Contrast with authentication.

Authorization Code - A multiple digit number or alphanumeric string entered to identify the user's authorization and/or level of authority for use of a system.

Autoexecute -The ability of an operating system to run certain programs without user intervention.

Automatic - Self-acting. Operating by its own mechanism, based on a non-personal stimulus.

Automatic Restart - A function in which a process can automatically restart from a point of failure. The halt may have been the result of a power or circuit failure or an interrupt generated by a user.

Average Power - The average over time of a modulated signal.

AWG - See American Wire Gauge.

Azimuth - Horizontal angle measured from true north.

B - Byte.

b - bit.

Babble - Crosstalk from multiple channels or circuits.

Back reflection, optical return loss - Light reflected from the cleaved or polished end of a fiber caused by the difference of refractive indices of air and glass. Typically 4 percent of the incident light. Expressed in dB relative to incident power.

Back-up (noun) - The copy of data that results from a back-up procedure. Sometimes refers to the storage media that contains that data.

Back-up (verb) - To copy a file, directory, or column onto another storage device so that the data is retrievable if the original source is accidentally corrupted or destroyed.

Backboard - A wooden (or metal) panel used for mounting equipment usually on a wall.

Backbone - The main connectivity device of a distributed system. All systems that have connectivity to the backbone will connect to each other. This does not stop systems from setting up private arrangements with each other to bypass the backbone for cost, performance or security.

Backbone Cable - A main Cable run vertically (or horizontally) in a building to provide wire connectivity to separate areas in the building. It is not designed for direct system access.

Backbone Closet - Space provided in a building for terminating pairs of wire and connecting the backbone cable to systems in that area. See also: Telecommunications Closet.

Backbone Wiring - See backbone cable.

Backfeed Pull - A cable pulling method that starts in the middle of a conduit and feeds the cable, in one direction and then the other, from that location.

Background Noise - Extra signals that are found on a circuit, line or channel.

Backscattering - The return of a portion of scattered light to the input end of a fiber; the scattering of light in the direction opposite to its original propagation.

Backup Link - An alternate circuit that is not used until, or if, the primary link fails.

Ballast - An electrical circuit component used with fluorescent lamps to provide the voltage necessary to strike the mercury arc within the lamp, and then to limit the amount of current that flows through the lamp.

Balun - Balanced/unbalanced. Refers to an impedance-matching device used to connect balanced twisted-pair cabling with unbalanced coaxial cable, best known in the IBM cabling system.

Band Pass Filter - An electronic device that filters out all signals except those one or more selected frequency ranges.

Bandwidth - Technically, the difference, in Hertz (Hz), between the highest and lowest frequencies of a transmission channel. Usually identifies the capacity or amount of data that can be sent through a given circuit.

BASIC - Beginner's All-purpose Instruction Code. A computer compiler language; similar to FORTRAN.

Basic Input/Output System (BIOS) - A set of programs, usually in firmware, that lets each computer's central processing unit communicate with printers, disks, keyboards, consoles, and other attached input and output devices.

Battery - A device that converts chemical energy into electrical current.

Battery cycle life - The number of cycles that a battery can undergo before failing.

Battery self-discharge - Loss of chemical energy in a battery that is not under load.

Baud - A unit of signaling speed. The speed in Baud is the number of discrete conditions or signal elements per second. If each signal event represents only on bit condition, then Baud is the same as bits per second. Baud rarely equals bits per second.

Baud Rate - The rate at which data is transferred over an asynchronous RS-232 serial connection.

BBS - See: Bulletin Board System.

Beamsplitter - An optical device, such as a partially reflecting mirror, that splits a beam of light into two or more beams and that can be used in fiber optics for directional couplers.

Bearing Wall - A wall supporting a load other than its own weight such as the next floor above. Also called a load bearing wall.

Bend loss - A form of increased attenuation in a fiber that results from bending a fiber around a restrictive curvature (a macrobend) or from minute distortions in the fiber (microbends).

Bend Radius - the radius a cable can bend without risking breakage or increasing attenuation.

BER - Basic Encoding Rules. Standard rules for encoding data units described in ASN.1. Sometimes incorrectly lumped under the term ASN.1, which properly refers only to the abstract syntax description language, not the encoding technique.

Beta Test - The stage at which a new product is tested under actual conditions. These tests are usually performed by selected product users. See also Alpha Test.

Binary - The base-2 number system using only the symbols 0 and 1. since 0 and 1 can be represented as on and off, or negative and positive charges, most computers do their calculations in binary.

BIOS - See Basic Input/Output System.

Bit - A binary digit; must be either a 0 or a 1. It is the smallest unit of information and indicates one of the two electrical states: off (O) or on (1) in a computer.

Bit Error Rate (ber) - In testing, the ratio between the total number of bits transmitted in a given message and the number of bits in that message received in error. A measure of the quality of a data transmission, usually expressed as a number referred to a power of 10; e.g., 1 in 10 over 5.

Bit Rate - See bits per second.

Bit-mapped - Refers to a display screen on which a character or image is generated and refreshed according to a binary matrix (bit map) at a specific location in memory.

Bits Per Second (bps) - Basic unit of measurement for serial data transmission capacity, abbreviated as k bps, or kilobit/s, for thousands of bits per second; m bps, or megabit/s, for millions of bits per second; g bits, or gigabit/s for billions of bits per second; t bps, or terabit/s, trillions of bits per second.

Black Box - Any electronic device that is not understood. Also used to describe a device who's detailed components or functions can be ignored in a discussion.

Block - A unit of stored or transmitted data. It is usually in a standard size such as 512, 1024 or 4,096 bytes of data.

Blocking diode - A diode used to prevent current flow in a photovoltaic array during times when the array is not producing electricity.

Bonding - A very-low impedance path accomplished by permanently joining non-current-carrying metal parts. It is done to provide electrical continuity and to offer the capacity to safely conduct any current.

Bonding jumper - A conductor used to assure the required electrical connection between metal parts of an electrical system.

Bonding Conductor - The conductor that connects the noncurrent- carrying parts of electrical equipment, cable raceways, or other enclosures to the approved system ground conductor.

Boot, or bootstrap (computer) - A short series of instruction codes that program a computer to read other codes. A machine that has no program at all in its memory cannot even read data.

Boot up - See boot.

Braid - A woven group of filaments that covers one or more insulated conductors, particularly in coax cabling.

Branch circuit - Conductors between the last overcurrent device and the outlets.

Branch circuit, general purpose - A branch circuit that supplies outlets for lighting and power.

Branch circuit, individual - A branch circuit that supplies only one piece of equipment.

Branch circuit, multiwire - A branch circuit having two or more ungrounded circuit conductors, each having a voltage difference between them, and a grounded circuit conductor (neutral) having an equal voltage difference between it and each ungrounded conductor.

Breakout Box - A device that allows access to individual points on a physical interface connector for testing and monitoring. it is often used to troubleshoot RS-232 circuits and cables.

Breakout Cable - A multifiber cable where each fiber is protected by an additional jacket and strength element beyond that provide for the overall cable.

Broadband - A transmission medium capable of supporting a wider range of frequencies than that required for a single communication channel. It can simultaneously carry multiple signals by dividing the total capacity of the medium into multiple, independent bandwidth channels, where each channel operates only on specific range of frequencies. Broadband implies the use of a frequency agile modem to select the correct channel rather than direct modulation as used in baseband. See also: baseband.

Buffer - In systems, a portion of memory designated for temporary storage of data. In cabling, a protective material on the fiber coating to protect the fiber-optic cable. See also tight buffer or loose buffer.

Building - A structure which is either standing alone, or cut off from other structures by fire walls.

Building Core - That portion of any building devoted to elevators, stairwells, vertical plumbing, rest rooms, vertical electrical and communications cables and equipment.

Bulletin Board System (BBS) - A computer, and related software, which typically provides electronic messaging services, archives of files, and any other services or activities of interest to the bulletin board systems's operator.

Bus Bar - The heavy copper or aluminum bar used to carry currents in switchboards.

Bus (data) - The primary signal route, inside a computer, which can have several devices connected, letting them transmit and receive data at the same time. For example, IBM's Micro Channel Architecture.

Bus (network) - The main (multiple access) network cable or line that connects network stations. Also refers to a network topology of multiple stations communicating directly with the same cable with terminators at both ends, like an Ethernet or token bus.

Busy Tone - A single tone that is interrupted at 60 ipm (impulses per minute) to indicate that the terminal point of a call is already in use.

Bypass Diode - A diode connected in parallel with a block of parallel modules to provide an alternate current path in case of module shading or failure.

Byte - One character of information, usually 8-bits.

Cable plant, fiber optic - The combination of fiber optic cable sections, connectors, and splices forming the optical path between two terminal devices.

Cable Assembly - A completed cable, with its connectors and hardware, that is ready for installation.

Cable Patch Panel - A passive device frequently located in the intermediate distribution facility (IDF) or satellite equipment room to offer easy circuit cross connections. It is used to connect two sets of wire (e.g., the wire from the IDF to the office and the wire between the IDFs).

Cable System - All the cables and devices used to interconnect stations; often called the premises network.

Cache - A portion of a computer's RAM reserved to act as a temporary memory for items read from a disk. They become instantly available to the user.

CAD/CAM - Computer aided design/computer aided manufacture. Software/hardware combinations for the automation of engineering environments.

Campus Backbone - The primary wiring that travels through a campus. Buildings and their internal networks connect to the backbone for communications with each other.

Capacitor - An electrical device which causes the current in a circuit to lead the voltage, the opposite effect of induction.

Carrier Frequency - Frequency of the carrier wave that is modulated to transmit signals.

Category Cable - Cable that complies with EIA/TIA TSB 36 and is rated category 1 thru 5. The higher the category number, the better the cable will be at carrying high speed signals. Category wire may be shielded, but is always 100 ohm.

Category Devices - Devices, such as patch panels and wall jacks, that comply with EIA/TIA TSB40. See category cable.

Cathode - The negative electrode in an electrochemical cell.

CATV - An abbreviation for community antenna television or cable TV.

Central Office - The site where communications common carriers (telephone companies) terminate customer lines and house the equipment that interconnects these lines.

Character Set - A collection of characters, such as ASCII or EBCDIC, used to represent data in a system. These character are typically available on a keyboard or through a printer.

Charge controller - A device that controls the charging rate and state of charge for batteries. See Charge Rate.

Charge Rate - The rate at which a battery is recharged. Expressed as a ratio of battery capacity to charge current flow, for instance, C/5.

Charge controller - A device that controls the charging rate and state of charge for batteries. See Charge Rate.

Charge Rate - The rate at which a battery is recharged. Expressed as a ratio of battery capacity to charge current flow, for instance, C/5.

Chromatic dispersion - The temporal spreading of a pulse in an optical waveguide caused by the wavelength dependence of the velocities of light.

Circuit breaker - A device used to open and close a circuit by automatic means when a predetermined level of current flows through it.

Circuit Switching - A communications method in which a dedicated path is identified by switching a signal to the wires that will connect the two hosts. The telephone system is an example of a circuit switched network. See also: connection-oriented, connectionless, packet switching.

Cladding - The outer concentric layer that surrounds the fiber core and has a lower index of refraction.

Clock - The timing signals used in data communications or the source of those timing signals.

CMOS - Complimentary Metal Oxide Semiconductor.

CO - See Central Office. Location where communications common carriers terminate customer lines and house the equipment that interconnects these lines.

Co-Processor - An additional central logic unit which performs specific tasks while the main unit executes its primary tasks. Frequently, these chips are added to speed up mathematical tasks or perform I/O functions.

Coaxial Cable - A transmission medium noted for its wide bandwidth and for its low susceptibility to interference. It is made up of an outer woven conductor which surrounds the inner conductor. the conductors are commonly separated by a solid insulating material.

COBOL - Common business-oriented language. One of the first standardized computing languages.

Code - (verb) To write compilable software. (noun) The compilable software that was written. Also the rules specifying how data may be represented in a particular system. Slang for program.

Common mode - Placed upon both sides of an amplifier at the same time.

Common Carrier - An organization in the business of providing regulated telephone, telegraph, telex, and data communications services.

Communications Protocol - The rules used to control the orderly exchange of information between stations on a data link or on a data network or system. Also called line discipline or protocol.

Communications Satellite - An earth satellite designed to act as a communications radio relay and usually positioned in geosynchronous orbit 35,800 kilometers (23,000 miles) above the equator so that it seems to be stationary in space.

Component - A type of network element like routers, computers, operating systems, gateways, etc.

Compression - A method to reduce the number of bits required to represent data.

Computer Network - An interconnection of computer systems, terminals, communications facilities, and data collecting devices.

Concealed - Made inaccessible by the structure or finish of the building.

Concentrator - 1) A photovoltaic module that uses optical elements to increase the amount of sunlight incident on a PV cell. 2) A communications device that offers the ability to concentrate many lower-speed channels into and out of one or more high-speed channels.

Conductor - A substance which offers little resistance to the flow of electrical currents. Insulated copper wire is the most common form of conductor.

Conduit body - The part of a conduit system, at the junction of two or more sections of the system, that allows access through a removable cover. Most commonly known as condulets, LBs, LLs, and LRs.

Continuous load - A load whose maximum current continues for three hours or more.

Conduit body - The part of a conduit system, at the junction of two or more sections of the system, that allows access through a removable cover. Most commonly known as conduits, LBs, LLs, LRs, etc.

Conferencing - A term used for communication software that allows participants to post notes. It is unlike electronic mail since participants do not have to be explicitly addressed. A primary function of a bulletin board. It also lets multiple users simultaneously interact online as a computer conference call.

Configuration - Settings that control the way a system or service will operate. Also the combined services and/or equipment that make up a communications system.

Connect Time - A measure of system usage. It is the interval during which the user was online for a session.

Continuous load - A load whose maximum current continues for three hours or more.

Controller - A device or group of devices that control (in a predetermined way) power to a piece of equipment.

Conversion Efficiency - The ratio of the electrical energy produced by a photovoltaic cell to the solar energy received by the cell.

Copper Distributed Data Interface (CDDI) - A variation of FDDI that uses Category 5 unshielded twisted pair copper wire. See also FDDI, TP-PMD.

Core - The central, light-carrying part of an optical fiber; it has an index of refraction higher than that of the surrounding cladding.

Core storage - Bianary memory storage, made up of tiny magneticelements.

CPU - Central Processing Unit. See also microprocessor.

Cross Connection - The wire connections running between wiring terminal blocks. Cross connections let pairs in one cable connect to pairs in another cable. See also patch panel, on the two sides of a distribution frame, or between binding posts in a terminal.

Cross-sectional area - The area (in square inches or circular mils) that would be exposed by cutting a cross-section of the material.

Crosstalk - The unwanted energy transferred from one circuit or wire to another which interferes with the desired signal. Usually caused by excessive inductance in a circuit.

CRT - Cathode Ray Tube (also generic reference to a terminal).

Crystalline Silicon - A type of PV cell made from a single crystal or polycrystalline slice of silicon.

Current - The flow of electricity in a circuit, measured in amperes.

Cursor - A movable underline, rectangular-shaped block of light, or an alternating block of reversed video on the screen of a display device, usually indicting where the next character is to be entered.

Customer Access Line Charge (CALC) - The FCC-imposed monthly surcharge added to all local lines to recover a portion of the cost of telephone poles, wires, etc., from end users. Before deregulation, a large part of these costs were financed by long distance users in the form of higher charges.

Customer Premise Equipment (CPE) - Equipment, usually including wiring located within the customer's part of a building.

Cut-out box - A surface mounted electrical enclosure with a hinged door.

Cutback method - A technique for measuring the loss of bare fiber by measuring the optical power transmitted through a long length then cutting back to the source and measuring the initial coupled power.

Cutoff wavelength - The wavelength beyond which single-mode fiber only supports one mode of propagation.

Cutoff Voltage - The voltage at which the charge controller disconnects the array from the battery. See Charge Controller.

Cyberspace - A term to describe the "world" of computers, and the society that gathers around them.

Daisy chaining - The connection of multiple devices in a serial fashion. Daisy chaining can save on transmission facilities. If a device malfunctions all of the devices daisy chained behind it are disabled.

Data rate - The number of bits of information in a transmission system, expressed in bits per second (bps), and which may or may not be equal to the signal or baud rate.

Data Base - A large, ordered collection of information.

Data Communications - The interchange of data messages from various sources are accumulated.

Data Communications Equipment - The equipment which provides the functions of interfacing between data terminal equipment and a communications channel DCE is normally a modem.

Data Link - A transmission path directly connecting two or more stations (a station may be a terminal, terminal controller, front end processor, or other type of digital equipment).

Data Terminal - A station in a system capable of sending and/or receiving data signals.

Data Transmission - The sending of data from one place for reception elsewhere.

dB -Decibel referenced to a microwatt.

dBm - Decibel referenced to a milliwatt.

Decibel - 1) A standard logarithmic unit for the ratio of two powers, voltages, or currents. In fiber optics, the ratio is power. 2) Unit for measuring relative strength of a signal parameter such as power or voltage.

$$dB = 10 \log^{10} \left(\frac{P^1}{P^2} \right)$$

Decimal - a digital system that has ten states, 0 through 9.

Deep Cycle - Battery type that can be discharged to a large fraction of capacity. See Depth of Discharge.

Default Route - A routing table entry which is used to direct any data addressed to any network numbers not otherwise listed in the routing table.

Degauss - To remove residual permanent magnetism.

Depth of Discharge (DOD) - The percent of the rated battery capacity that has been withdrawn.

Detector - An optoelectronic transducer used in fiber optics for converting optical power to electric current. In fiber optics, usually a photodiode.

Device (Also used as wiring device) - The part of an electrical system that is designed to carry, but not use, electrical energy.

Diagnostics - Programs or procedures used to test a piece of equipment, a communications link or network.

Dial Tone - A tone indicating that automatic switching equipment is ready to receive dial signals.

Diameter-mismatch loss - The loss of power at a joint that occurs when the transmitting half has a diameter greater than the diameter of the receiving half. The loss occurs when coupling light from a source to fiber, from fiber to fiber, or from fiber to detector.

Diffuse Radiation - Radiation received from the sun after reflection and scattering by the atmosphere.

Digital - Communications procedures, techniques, and equipment whereby information is encoded as either binary "1" or "0". Also the representation of information in discrete binary form.

Diode - Electronic component that allows current flow in one direction only.

Dip Switch - A dual in-line package switch. It has two parallel rows of contacts that let the user switch electrical current through a pair of those contacts to on or off. They are used to reconfigure computer components and peripherals. See also jumpers.

Direct current (DC) - Electrical current which flows in one direction only.

Disconnecting means - A device which disconnects a group of conductors from their source of supply.

Discrete Access - An access method used in star LANs: each station has separate (discrete) connections it uses for the LAN's switching capability. Contrast with shared access.

Disk - A rotating disk covered with magnetic material, used for storage of data.

Disk Operating System - A program or set of programs that tells a disk-based computer system to schedule and supervise work, manage computer resources, and operate and control its peripheral devices.

Disk Server - A LAN device that lets multiple users access sections of its disks for creating and storing files. Contrast with a file server, which allows users to share files.

Dispersion - A general term for those phenomena that cause a broadening or spreading of light as it propagates through an optical fiber. The three types are modal, material, and wave-guide.

Distortion - An unwanted change to signal caused by outside interference or by imperfections of the transmission system. Often caused by excess capacitance.

Distributed Data Processing - The processing of information in separate locations as on a local area network. This is a more efficient use of processing power since each CPU can do a certain task.

Distributed Database - A collection of several different data repositories that looks like a single database to the user.

Distribution Block or Frame - Centralized equipment where wiring is terminated and cross-connections are made. See also Intermediate Distribution Frame and Main Distribution Frame.

Domain - A zone or part of a naming hierarchy. An Internet domain name consists of a sequence of names separated by periods, as in "roadie.cs.arg3.com."

DOS - See Disk Operating System.

Down Time - The total time a system is out of service due to equipment failure.

Download - To make a copy of a file from a central service (or server) onto a local computer. Contrast with a load and execute operation on a LAN.

DRAM - Dynamic Random Access Memory. Contrast with Static RAM.

Driver - The amplifier stage preceding the output stage.

Drop (noun) - A portion of cable that connects a user station to a network. Also refers to the jack that is the point of contact for the cable drop. The user sees a drop as a network connection.

Drop (verb) - To logically disconnect part or all of a signal whether intentionally or unintentionally.

Drop Cable - Cable that provides access to and from a network system. Possibly the cable from a transceiver or an individual line in a multi-drop situation. Also the cable from a wall-mounted faceplate or jack to a user's system.

Dumb Terminal - A term used to describe an asynchronous, ASCII terminal that, although it may be "intelligent" in many of the functions it provides, uses no communication protocol. It may operate at speeds up to 19.2 Kbps.

Duplex cable - A two-fiber cable suitable for duplex transmission.

Duplex transmission - Transmission in both directions, either one direction at a time (half duplex) or both directions simultaneously (full duplex).

Duplex - Simultaneous two-way independent transmission.

Earth Station - Ground-based equipment used to communicate via satellites.

Edge-emitting diode (E-LED) - A LED that emits from the edge of the semiconductor chip, producing higher power and narrower spectral width.

EDP - Electronic data processing.

EEPROM - See Electrically Erasable Programmable Read-Only Memory.

EIA - See Electronics Industries Association.

EIA/TIA 568 - The commercial building wiring standard. It defines a generic wiring system for a multi-product, multi-vendor environment. Widely considered the most important standard for building wiring.

EIA/TIA 570 - The residential and light commercial building wiring standard. It is the wiring standard for single and multi-family residential and mixed use facilities.

EIA-232 - Standard interface definition for serial devices. better known as RS-232. RS-232-E is the current version of the standard.

Electrically Erasable Programmable Read-Only Memory (EEPROM) - A memory that can be electronically programmed and erased, but which does not need a power source to hold the data.

Electrolyte - A liquid or paste in which the conduction of electricity is by a flow of ions.

Electromagnetic Interference (EMI) - The energy given off by electronic circuits and picked up by other circuits; based on the type of device and operating frequency. EMI effects can be reduced by shielding and other cable designs. Minimum acceptable levels are detailed by the FCC. See also Radio Frequency Interference.

Electronic Data Interchange (EDI) - A standard system of exchanging order and billing information between computers in different companies.

Electronic Mail (email) - A system that lets computer users exchange messages with other computer users (or groups of users). The messages go to an email server instead of directly to the end recipient.

Electronic Switch - A modern programmable switch (also called ESS, for Electronic Switching System). It contains only solid state electronics, unlike older mechanical switches.

Electronics Industries Association (EIA) - A US trade organization that issues its own standards and contributes to ANSI. Best known for its development RS-232 and the building wiring standard, 568. Membership includes US manufacturers.

Email - See Electronic Mail.

Email Address - The address that is used to send electronic mail to a specific destination. For example: gsimen@netcom.com.

EMI - See Electromagnetic Interference.

EMP - Electromagnetic Pulse.

Enclosed - Surrounded by a case, housing, fence, or walls that prevent unauthorized people from contacting the equipment.

Encryption - Manipulation of a packet's data in order to prevent any but the intended recipient from reading that data.

End User - The human source and/or destination of information sent through the communications system.

EOM - End of Message.

EPROM - Erasable Programmable Read Only Memory.

Equalization - The process of restoring all cells in a battery to an equal state of charge.

ESCON - IBM standard for connecting peripherals to a computer over fiber optics. Acronym for Enterprise System Connection.

Ethernet - A 10-Mbps, coaxial standard for LANs, initially developed by Xerox and later refined by Digital, Intel and Xerox (DIX). All nodes connect to the cable where they contend for access via CSMA/CD. Also slang for the coaxial cable that carries the standard.

Ethernet Controller - A device that controls a computer's access to Ethernet services. The CSMA/CD protocols are used by the controller to free the CPU.

Ethernet Meltdown - The result of an event that causes saturation, or near saturation, on an Ethernet. It usually comes from illegal or misrouted packets and normally runs for a short time only.

Event - The occurrence of a particular change in the state of a managed object.

Event Log - A record of significant events.

Excess loss - In a fiber-optic coupler, the optical loss from that portion of light that does not emerge from the nominally operational ports of the device.

Exchange - The collection of equipment in a communications system that controls the connection of incoming and outgoing lines. Also known as central office.

EXEC - Executable.

Exposed - Able to be inadvertently touched or approached.

Extrinsic loss - In a fiber interconnection, that portion of loss that is not intrinsic to the fiber but is related to imperfect joining, which may be caused by the connector or splice.

Facsimile - A copy. Also the transmission of documents via communications circuits. This is done by using a device which scans the original document, transforms the image into coded signals and reproduces documents it receives. Also called fax.

Fading - A situation, generally of microwave or radio transmission, where external influences cause a signal to be deflected or diverted away from the target receiver. Also the reduction in intensity of the power of a received signal.

Fall time - The time required for the trailing edge of a pulse to fall from 90% to 10% of its amplitude; the time required for a component to produce such a result. "Turnoff time". Sometimes measured between the 80% and 20% points.

FAQ - Frequently Asked Questions.

Farad - The unit of measurement of capacitance.

Fault - A condition that causes any physical component of a system to fail to perform in acceptable fashion.

Fault Tolerance - The ability of a program or system to operate properly even if a failure occurs.

FDDI - See Fiber Distributed Data Interface.

FDDI II - See Fiber Distributed Data Interface.

Federal Communications Commission (FCC) - The government agency established by the Communications Act of 1934 which regulates interstate communications.

Feeder - Circuit conductors between the service and the final branch circuit over current device.

Ferrule - A precision tube that holds a fiber for alignment for interconnection or termination. A ferrule may be part of a connector or mechanical splice.

Fiber identifier - A device that clamps onto a fiber and couples light from the fiber by bending, to identify the fiber and detect high-speed traffic of an operating link or a 2 kHz tone injected by a test source.

Fiber tracer - An instrument that couples visible light into the fiber to allow visual checking of continuity and tracing for correct connections.

Fiber Channel - A switched datalink technology released as a draft by ANSI. It uses coaxial cable and shielded twisted pair at speeds of 133 Mbps to 1 Gbps.

Fiber Distributed Data Interface (FDDI) - A high-speed (100Mb/s) LAN standard. The underlying medium is fiber optics, and the topology is a dual-attached, counter-rotating token rings,. FDDI II is a draft standard for revising FDDI to carry multi-media signals at 100 Mbps. See also: Local Area Network, token ring.

Fiber Optics - A technology that uses light as a digital information carrier. Fiber optic cables are direct replacement for conventional cables and wire pairs. They occupy far less physical space and are immune to electrical interference.

File (computer) - A block of data.

File Server - In local networks, a station dedicated to providing file and mass data storage services to the other network stations.

File Transfer - The copying of a file from one computer to another over a network or dial-up circuit.

File Transfer Protocol (FTP) - A TCP/IP protocol that lets a user on one computer access, and transfer files to and from, another computer over a network. FTP is usually the name of the program the user invokes to accomplish this task.

Filter - A combination of circuit elements which is specifically designed to pass certain frequencies and resist all others.

Firmware - Permanent or semi-permanent micro-instruction control for a user-oriented function.

Flame - A strong opinion and/or criticism of something, usually a frank inflammatory statement, in an electronic mail message. It is common to precede a flame with an sign of pending fire (as in "FLAME ON!").

Float Charge - The charge to a battery having a current equal to or slightly greater than the self discharge rate.

Floating Point - A native data type on most operating systems that can have numbers after the decimal point. Contrast with an integer.

Fluorescence - The emission of light by a substance when exposed to radiation or the impact of particles. The effect ceases in a fraction of a second once the source of radiation or particles is removed.

FM - Frequency modulation.

FO - common abbreviation for fiber optic.

Footprint - (1) The space a device occupies on a desk or work surface. (2) The precise area of the earth in which a satellite communications signal can be received.

Four Wire Circuits - Telephone circuits which use two separate one-way transmission paths of two wires each, as opposed to regular local lines which usually only have two wires to carry conversations in both directions.

Four-Wire Channel - A circuit containing two pairs of wire (or their logical equivalent) for simultaneous (i.e., full-duplex) two-way transmission.

Frame Relay - A faster form of packet switching that is accomplished with smaller packet sizes and less error checking.

Frequency modulation - A method of transmission in which the carrier frequency varies in accordance with the signal.

Frequency - The number of times per second a signal regenerates itself at a peak amplitude. It can be expressed in hertz (Hz), kilohertz (Hz), megahertz (MHz), etc.

Frequency Division Multiplexing (FDM) - A method of dividing an available frequency range into subparts, each having enough bandwidth to carry one channel.

Fresnel reflection - The reflection that occurs at the planar junction of two materials having different refractive indices; Fresnel reflection is not a function of the angle of incidence.

Fresnel reflection loss - Loss of optical power due to Fresnel reflections.

Full Duplex - A circuit that lets messages flow in both direction at the same time. Contrast with half-duplex where only one side can transmit at a time.

Fusion splicer - An instrument that splices fibers by fusing or welding them, typically by electrical arc.

GAN - Global Area Network.

Gap loss - Loss resulting from the end separation of two axially aligned fibers.

Gassing - Gas by-products produced when charging a battery. Also, termed out-gassing.

Gateway - The original Internet term for a router or more accurately, an IP router. In current usage, "gateway" and "application gateway" refer to translating systems that convert data traveling from one environment to another.

GB - Gigabit (or gigabyte) backbone. The gigabit backbone is an effort to increase the speed of the Internet to one Gbps.

Gb - Gigabit. One billion bits of information.

Gbyte - Gigabyte. One billion bytes of data.

GFLOPS - Billion Floating Operations Per Second.

Giga - A prefix that means one billion.

Graded-index fiber - An optical fiber whose core has a nonuniform index of refraction. The core is composed of concentric rings of glass whose refractive indices decrease from the center axis. The purpose is to reduce modal dispersion and thereby increase fiber bandwidth.

Grid - Term used to describe an electrical utility distribution network.

Ground - An electrical connect (on purpose or accidental) between an item of equipment and the earth.

Hacker - A person who delights in having an intimate understanding of the internal workings of computers and computer networks.

Half Duplex - A circuit for transmitting or receiving signals in one direction at a time.

Hand Shaking - The exchange of predetermined control signals for establishing a session between data sets.

Hard copy (computer) - A printout on paper or cards.

Hardware (computer) - The physical computer and related machines.

Hardwire - To wire or cable directly between units of equipment.

Harmonic - 1) A sinusoid which has a frequency which is an integral multiple of a certain frequency. 2) The full multiple of a base frequency.

HDTV - High Definition Television.

Hertz (Hz) - International standard unit of frequency. Replaces the identical older "Cycles-per-second."

Hexadecimal - A number system with 16 members represented by 0 through 9 followed by A through F. Each character identifies four bit or a half-byte. Also called hex.

Host - (1) A computer that provides services directly to users, i.e. the user's computer. In TCP/IP, an IP addressed device. (2) A large computer that serve many users, i.e. a mini-computer or mainframe.

Host Address - The part of an Internet address that designates which node on the (sub)network is being addressed. Also called host number.

Host Computer - See host.

Hub - A device which connects to several other devices usually in a star topology. Also called: concentrator, multiport repeater or multi-station access unit (MAU).

Hybrid - An electronic circuit that uses different cable types to complete the circuit between systems.

Hz - See Hertz.

I/O or Input-Output - Related to the process of getting data into and out of a computer or processor.

Identified (for use) - Recognized as suitable for a certain purpose, usually by an independent agency, such as U. L.

IDP - Integrated detector/preamplifier.

IEEE - Institute of Electrical and Electronic Engineers (US).

Impedance - The effects placed upon an alternating current circuit by induction, capacitance, and resistance. Total resistance in an AC circuit.

Index of refraction - The ratio of the velocity of light in free space to the velocity of light in a given material. Symbolized by n.

Index-matching material - A material, used at optical interconnection, having a refractive index close to that of the fiber core and used to reduce Fresnel reflections.

Inductance - The characteristic of a circuit that determines how much voltage will be induced into it by a change in current of another circuit.

Insertion loss - The loss of power that results from inserting a component, such as a connector or splice, into a previously continuous path.

Integrated circuit or IC - A circuit in which devices such as transistors, capacitors, and resistors are made from a single piece of material and connected to form a circuit.

Interconnect - (1) The arrangement that allows the connection of customer's communications equipment to a common carrier network. (2) The generic term for a circuit administration point that allows routing and rerouting of signal traffic.

Interface - The point that two systems, with different characteristics, connect.

Internet - (Note the capital "I") The largest Internet in the world including large national backbone nets (such as MILNET, NSFNET, and CREN) and many regional and local networks world- wide. The Internet uses the TCP/IP suite. Networks with only email connectivity are not considered on the Internet.

Interoperability - The ability of software and hardware on multiple machines from multiple vendors to communicate meaningfully.

Intrastate - Any connection made that remains within the boundaries of a single state.

Inventor - A device for changing direct current into alternating current by alternately switching DC in inverted polarity.

Ion - An atom or molecule that has acquired a charge by gaining or losing one or more electrons.

IRQ - Interrupt Request

Isolated - Not accessible unless special means of access are used.

Isolation transformer - A one to one transformer that is used to isolate the equipment at the secondary from earth ground.

Jack - A receptacle (female) used with a plug (male) to make a connection to in-wall communications cabling or to a patch panel.

Jacket - The protective and insulating outer housing on a cable. Also called a sheath.

Jumper - Patch cable or wire used to establish a circuit, often temporarily, for testing or diagnostics. Also the devices, shorting blocks, used to connect adjacent exposed pins on a printed circuit board that control the functionality of the card.

Junction Box - A box, usually metal, that encloses cable connections for their protection.

K Band - A microwave band from 10.9 GHz to 36GHz.

KB - Kilobyte. One thousand bytes.

Kb - Kilobit. One thousand bits.

Kbps - Kilobits per second. Thousand bits per second.

Kbyte - Kilobyte. One thousand bytes.

Kermit - A popular file transfer protocol developed by Columbia University. By running in most operating environments, it provides an easy method of file transfer. Kermit is NOT the same as FTP. See also File Transfer Protocol.

KU Band - The frequency band from 12 to 14 GHz that is used for satellite communications.

kW - The abbreviation for kilowatt, a unit of measurement of electrical power. One kilowatt is equal to one thousand watts.

L Band - Microwave and satellite communications frequencies in the 390 MHz to 1550 MHz range.

Lamp - A light source. Reference is to a light bulb, rather than a table lamp.

LAN - See Local area network.

LAN Adapter - An external device or card, for internal use that lets a device gain access to a local area network.

Laser - See Light Amplitude by Stimulated Emission of Radiation.

Laser Optical - The generic term for the system of recording data, sound or video on optical discs.

Launch cable - A known good fiber optic jumper cable attached to a source and calibrated for output power used for loss testing. This cable must be made of fiber and connectors of a matching type to the cables to be tested.

Layer - A modular portion of a stacked protocol that consists of one or more semi-independent protocols. Each layer builds on the layer beneath it and feeds information to the protocols in the layers above it. TCP/IP has five layers of protocols, and OSI has seven.

LED - Light emitting diode.

Level - An expression of the relative signal strength at a point in a communications circuit as compared to a standard.

Life-Cycle Cost - The estimated cost of owning and operating a system for the period of its useful life.

Light Amplitude by Stimulated Emission of Radiation (laser) - A device that produces a very uniform, single frequency of light. Digital signals can be transmitted by turning it on and off rapidly. See also Fiber Optics.

Light Emitting Diode - A semiconductor diode that gives off light when current is passed through it.

Line - An electrical path between two points, usually a telco CO and the end user.

Line Driver - A short haul communications device for overcoming the RS-232 alleged distance limitation. See also short haul modem or limited distance modem.

Link - The physical interconnection between two systems (sometimes called nodes) in a network. A link may consist of a data communications circuit or a direct cable connection.

Load - The amount of electric power used by any electrical unit or appliance at any given moment.

Local Area Network (LAN) - A data network intended to serve an area of only a few square kilometers or less. Because the network is known to cover only a small area, optimizations can be made in the network signal protocols that permit higher data rates. See also Ethernet, Fiber Distributed Data Interface, token ring, Wide Area Network.

Local Loop - the local connection between the end user and the Class 5 central office.

Location, damp (damp location) - Partially protected locations, such as under canopies, roofed open porches, etc. Also, interior locations that are subject only to moderate degrees of moisture, such as basements, barns, etc.

Location, dry (dry location) - Areas that are not normally subject to water or dampness.

Location, wet (wet location) - Locations underground, in concrete slabs, where saturation occurs, or outdoors.

Login - To gain access to a computer or network by identifying the acceptable user name and passing the required authentication procedure(s).

Long Haul - circuits spanning considerable distances.

Loss budget - The amount of power lost in a fiber optic link. Often used in terms of the maximum amount of loss that can be tolerated by a given link.

Loss, optical - The amount of optical power lost as light is transmitted through fiber, splices, couplers, and the like.

Machine language - Programs or data that is in a form that is immediately useable by the computer, usually bianary.

Macintosh - A computer made by Apple Computer that is characterized by the graphical, intuitive user interface.

Main Distribution Frame (MDF) - The point where outside cables and backbone cables are cross connected. Usually the point at which all cables from intermediate distribution frames intersect.

Mainframe - A generic term of a large, multi-user, multi-tasking computer. Many use it to refer to IBM mainframe computers.

MAN - See Metropolitan Area Network

Mbps - Million bits per second.

Mbyte - Megabyte. Million bytes of information.

MDF - See Main Distribution Frame

Mechanical splice - A semipermanent connection between two fibers made with an alignment device and index matching fluid or adhesive.

Media - The plural of medium.

Media Interface Connector - The optical fiber connector which connects the fiber to the FDDI controller.

Medium - (1) Any substance that can be used for the propagation of signals, such as optical fiber, cable, wire, dielectric slab, water, air, or free space. (2) the material on which data is recorded; for example, magnetic type, diskette.

Metropolitan Area Network (MAN) - A data network intended to serve an area approximating that of a large city. See also Local Area Network, Switched Multimegabit Data Service, Wide Area Network.

Microcomputer - A small-scale programmable machine that processes information. It generally has a single chip as its central processing unit and includes storage and input/output facilities in the basic unit.

Micron (m) - A unit of measure, 10^{-6} m, used to measure wavelength of light.

Microprocessor - The control unit of a microcomputer that contains the logical elements for manipulating and performing arithmetical and logical operations on information.

Microscope, fiber optic inspection - A microscope used to inspect the end surface of a connector for flaws or contamination or a fiber for cleave quality.

Microwave - (1) Portion of the electromagnetic spectrum above about 760 MHz. (2) High-frequency transmission signals and equipment that employ microwave frequencies, including line-of- sight open-air microwave transmission and satellite communications.

MIPS - Million instructions per second. A measure of the speed of a CPU.

Misalignment loss - The loss of power resulting from angular misalignment, lateral displacement, and end separation.

Modal dispersion - Dispersion resulting from the different transit lengths of different propagating modes in a multimode optical fiber.

Mode - A single electromagnetic field pattern that travels in fiber.

Mode filter - A device that removes optical power in higher-order modes in fiber.

Mode - In guided-wave propagation, such as through a waveguide or optical fiber, a distribution of electromagnetic energy that satisfies Maxwell's equations and boundary conditions. Loosely, a possible path followed by light rays.

Modem - A device which connects a computer to a telephone line and sends data over those phone lines, normally using a modulated audio tone.

Modem - A device which modulates and demodulates signals. It provides an interface between digital terminals and analog circuits and equipment.

Modulation - The process by which the characteristic of one wave (the carrier) is modified by another wave (the signal). Examples include amplitude modulation (AM), frequency modulation (FM), and pulse-coded modulation (PCM).

Module - The smallest replaceable unit in a PV array. An integral encapsulated unit containing a number of PV cells.

Motherboard - A board containing a number of printed circuit sockets and serving as a backing.

Motherboard - The central card of a computer that accepts other printed circuit cards. So called due to its female connectors and hierarchy in the computer. IBM now calls this a planar board in their System/2 computers.

Mouse - A handheld input device, separate from a keyboard, that is moved on a surface to control the position of an indicator (cursor) on a display screen.

Multi-access - The capability that lets multiple users simultaneously communicate with a computer or a network.

Multimode fiber - A fiber with core diameter much larger than the wavelength of light transmitted that allows many modes of light to propagate. Commonly used with LED sources for lower-speed, short-distance links.

Multiplex - To combine multiple input signals into one for transmission over a single high-speed channel. Two methods are used: (1) frequency division, and (2) time division.

Multiplexer - A device that lets more than one signal be sent simultaneously over one physical circuit. At the receiving end, the circuit is demultiplexed into the same number of outputs. Also called a mux.

Multiplexing - The process by which two or more signals are transmitted over a single communications channel. Examples include time-division multiplexing and wavelength-division multiplexing.

Multitasking - Two or more program segments running in a computer at the same time.

MUX - See Multiplexer

NA - Numerical aperture.

Nanometer (nm) - A unit of measure, 10^{-9} m, used to measure the wavelength of light.

Near End Crosstalk (NEXT) - Interference that transfers from the transmit side of a circuit to the receive side of the same circuit. It is measured at the transmit or "near" end.

NEC - National Electrical Code, which contains safety guidelines for all types of electrical installations.

Netiquette - A pun on "etiquette"; proper behavior on a specific network.

Network - A computer network is a data communications system which interconnects computer systems at various different sites. A network may include any combination of LANs, MANs or WANs.

Network File Transfer - A procedure that lets a network user (1) copy a remote file, (2) /translate file attributes, and (3) access remote accounts, either interactively or through a program.

Network Meltdown - A state of complete network overload. The network equivalent of gridlock. See also broadcast storm.

Network Operating System (NOS) - The software that manages the relationships between network resources and users. While there are several parts and protocols in a NOS, it is usually available as a single product.

Network Operations Center (NOC) - The site for monitoring the operation of and directing the management and maintenance of a network.

Network Trunks - Circuits connecting switching centers.

Next - See Near End Crosstalk.

Node - An addressable device attached to a computer network. See also host, router, server, user.

NOS - See Network Operating System

Novell - Makers of NetWare software for networks and the dominant company in local area networking.

Numerical aperture - The "light-gathering ability" of a fiber, defining the maximum angle to the fiber axis at which light will be accepted and propagated through the fiber. NA - sin O, where O is the acceptance angle. NA is also used to describe the angular spread of light from a central axis, as in exiting a fiber, emitting from a source, or entering a detector.

OCR - Optical Character Recognition.

OEM - See Original equipment manufacturer.

Ohm - The unit of measurement of electrical resistance. One ohm of resistance will allow one ampere of current to flow through a pressure of one volt.

Open Circuit Voltage - The maximum voltage produced by a photovoltaic cell, module, or array without a load applied.

Optical amplifier - A device that amplifies light without converting it to an electrical signal.

Optical loss test set (OLTS) - A measurement instrument for optical loss that includes both a meter and source.

Optical power - The amount of radiant energy per unit time, expressed in linear units of Watts or on a logarithmic scale, in dBm (where 0 dB = 1 mW) or dB (where 0 dB = 1 W).

Optical switch - A device that routes an optical signal from one or more input ports to one or more output ports.

Optical time-domain reflectometry - A method of evaluating optical fibers based on detecting backscattered (reflected) light. Used to measure fiber attenuation, evaluate splice and connector joints, and locate faults.

Optical Character Recognition (OCR) - a process than scans images, detects character shape matches, and converts them into digital code for the matching character.

Optical Disks - Storage devices that store data by using laser technology to record data. They Feature greater storage capacity than magnetic disks but currently offer slower access.

OS/2 - Operating System/2. Graphical successor to DOS for the IBM PC. It was developed jointly by Microsoft and IBM but now sold only by IBM.

OSI Reference Model - A seven-layer structure designed to describe computer network architectures and the way that data passes through them.

OTDR - Optical time-domain reflectometry.

Out-of-band - Any frequency separate from the band being used for data, voice or video traffic. Typically requires a completely separate signal path or wire. Contrast with in-band.

Outlet - The place in the wiring system where the current is taken to supply equipment.

Overcurrent - Too much current.

PABX - Private Automatic Branch Exchange

Packet Switch - A device to accept, route and forward packets in a packet switched network.

Packet Switching - A communications paradigm in which packets (messages) are individually routed between hosts, with no previously established communication path.

Parity - Having a constant state or equal value. Parity checking is one of the oldest error checking techniques.

Parity Bit - A check bit appended to an array of binary digits for parity checking.

PBX - Private branch exchange. A telephone switch which is installed at the customer premises.

PC - See Personal Computer.

PCS - Plastic-clad silica.

Personal Computer (PC) - (1) A desktop computer developed by IBM or a clone developed by a third-party vendor. (2) Sometimes used more generically to refer to other desktop systems, such as the Apple Macintosh. (3) The original, IBM developed, computer using an Intel 8088 cpu and an 8 bit internal bus.

Phase converter - A device that derives three phase power from single phase power. Used extensively is areas (often rural areas) where only single phase power is available, to run three phase equipment.

Photodector - An optoelectronic transducer, such as a pin photodiode or avalanche photodiode.

Photodiode - A semiconductor diode that produces current in response to incident optical power and used as a detector in fiber optics.

Photon - A quantum of electromagnetic energy. A "particle" of light.

Photovoltaic - Changing light into electricity.

Physical Layer - the OSI layer (layer 1) specification that defines signal voltages, encoding schemes and physical connections for sending bits across a physical media.

Physical Media - Any means in the physical world for transferring signals between systems. Since it is outside (below) the OSI Model, it is also know as "Layer 0". It includes all copper media, fiber optic media and wireless technologies.

Physical Unit (PU) - An SNA term used to refer to different types of hardware in the network.

Pigtail - A short length of fiber permanently attached to a component, such as a source, detector, or coupler.

Pixel - Picture Element. A single dot on a screen, or printout, that is represented by, or represents, a specific memory address.

Plain Old Telephone Service (POTS) - Standard, two-wire, telephone service. Contrast with ISDN, Call Waiting, or any other additional features.

Plastic fiber - An optical fiber having a plastic core and plastic cladding.

Plastic-clad silica fiber - An optical fiber having a glass core and plastic cladding.

Plenum cable - A cable whose flammability and smoke characteristics allow it to be routed in a plenum area without being enclosed in a conduit.

Plenum - A chamber which forms part of a building's air distribution system, to which connect one or more ducts. Frequently areas over suspended ceilings or under raised floors are used as plenums.

Point to Point - A network configuration that has a connection between only two, terminal installations as opposed to multipoint.

Point-to-Point Protocol (PPP) - A protocol for transmitting packets over serial, synchronous and asynchronous point-to-point circuits.

Polling - The process of inviting another station or node to transmit data. It requires that one station control the other stations.

POTS - See Plain Old Telephone Service.

Power budget - The difference (in dB) between the transmitted optical power (in dBm) and the receiver sensitivity (in dBm).

Power meter, fiber optic - An instrument that measures optical power emanating from the end of a fiber.

Power (Watts) - A basic unit of electrical energy, measured in watts. See also Watts.

Premises Distribution System - An AT & T cabling system that has also become a generic term for any structured cabling system.

Private Branch Exchange (PBX) - A private phone system (switch) that connects to the public telephone network and offers in-house connectivity. To reach an outside line, user must dial a digit like 8 or 9.

Private Line (PL) - A full-time, leased line that connects two points.

Process (verb) - To perform one or more operations on information.

Process (noun) - An operation performed on data or information.

Processor - A computer, or chip, capable of receiving information, manipulating it, and supplying results.

Program - A group of instructions that direct a computer's tasks.

Programmer - A person who designs, writes, and tests computer programs.

PROM - Programmable Read-Only Memory. A chip-based information storage area that can be recorded by an operator but only erased through a physical process.

Prompting - Messages from a computer that give instructions to the user.

Propagation Velocity - The speed that a signal travels through a medium from source to target. Also called Velocity of Propagation.

Protocol - A formal description of message formats and the rules computers must follow to exchange those messages.

Protocol Converter - A program or device that translates protocol in both directions.

Public Domain (PD) - Intellectual property that is freely available to all people.

Pulse spreading - The dispersion of an optical signal with time as it propagates through an optical fiber.

Punch Down - A wire termination technique in which the wire is laid on a connector and "punched" into place with a special tool. This procedure strips that portion of the insulation to provide an excellent, long lasting contact.

Radio Frequency Interference (RFI) - The unintentional transmission of radio signals. Computer equipment and wiring can both generate and receive RFI. See also electromagnetic interference.

Radius (Radii, plural) - The distance from the center of a circle to its outer edge.

RAM - See Random access memory.

Random Access Memory - Dynamic memory, sometimes known as main memory or core. See also DRAM.

RBOC - See Regional Bell Operating Company.

Receive cable - A known good fiber optic jumper cable attached to a power meter used for loss testing. This cable must be made of fiber and connectors of a matching type to the cables to be tested.

Redundancy - (1) The portion of the total information contained in a message that can be eliminated without loss of essential information. (2) Providing duplicate, devices to immediately take over the function of equipment that fails.

Regeneration - The process of receiving distorted signals and recreating them at the correct rate, amplitude, and pulse width.

Regional Bell Operating Company (RBOC) - One of seven regional holding companies that were the result of the AT & T divestiture (breakup).

Remote Site - Site not serviced by an electrical utility grid.

Remote Access - The ability of network nodes or remote computers to gain access to a computer which is at a different location.

Repeater - A device that receives, amplifies (and perhaps reshapes), and retransmits a signal. It is used to boost signal levels when the distance between repeaters is so great that the received signal would otherwise be too attenuated to be properly received.

Resistance - The opposition to the flow of current in an electrical circuit.

Resonance - A condition in an electrical circuit, where the frequency of an externally applied force equals the natural tendency of the circuit.

Response Time - The amount of time elapsed between a generation of an inquiry at a system and receipt of a response at that same system.

RG-58 - The coaxial cable used by Thin Ethernet (10 base 2). It has a 50 ohm impedance and so must use 50 ohm terminators.

RG-59 - The coaxial cable, with 75 ohm impedance, used in cable TV and other video environments.

RG-62 - the coaxial cable, with 93 ohm impedance, used by ARCNet and IBM 3270 terminal environments.

Ring network - A network topology in which terminals are connected in a point-to-point serial fashion in an unbroken circular configuration.

RISC - Reduced instruction set computer.

Rise time - the time required for the leading edge of a pulse to rise from 10% to 90% of its amplitude; the time required for a component to produce such a result. "Turnon time." Sometimes measured between the 20% and 80% points.

RJ - Registered Jack

RJ11 - A standard six conductor modular jack or plug that uses two to six conductors. Commonly used for telephones and some data communications.

RJ22 - The standard four conductor modular jack that connects a telephone handset to its base unit.

RJ45 - A standard eight conductor modular jack or plug that uses two to eight conductors. Replacing RJ11 for use with data communications and increasing use with telephones. The wire may be twisted or flat though flat will only work up to 19.2 Kbps.

ROM - Read-Only Memory

Root Directory - The top of a namespace or file system. Usually represented by a slash (\ in DOS, / in UNIX).

Rotary convertor - A type of phase convertor.

Route - (noun) The path that network traffic follows from its source host to its target host. (verb) To send a packet or frame of data through a network to its correct destination. In other words, what routers do.

RS232-C - An EIA physical interface standard for use between data communications equipment (DCE) and data terminal equipment (DTE). The current standard is RS-232-E

Satellite Relay - An active or passive repeater in geosynchronous orbit around the Earth which amplifies the signal it receives before transmitting it back to earth.

SCSI - See Small Computer Systems Interface. Pronounced "scuzz-ee".

Semiconductor - A material that has electrical characteristics somewhere in between those of conductors and insulators.

Sensitivity - For a fiber-optic receiver, the minimum optical power required to achieve a specified level of performance, such as a BER.

Separately derived system - A system whose power is derived (or taken) from a generator, transformer or convertor.

Serial (computer) - Handling data sequentially, rather than simultaneously.

Service - Equipment and conductors that bring electricity from the supply system to the wiring system of the building being serviced.

Service drop - Overhead conductors from the last pole to the building being served.

Shelf Life - The period of time that a device can be stored and still retain a specified performance.

Signal Circuit - An electrical circuit which supplies energy to one or more appliances which give a recognizable signal.

Signal to Noise Ratio - Ratio of the signal power to the noise power in a specified band, usually expressed in decibels.

Simplex - Simplex mode. Operation of a channel in one direction only with no capability of reversing.

Simplex cable - A term sometimes used for a single-fiber cable.

Simplex transmission - Transmission in one direction only.

Sine Wave - A waveform corresponding to a single-frequency, periodic oscillation, which can be shown as a function of amplitude against angle and in which the value of the curve at any point is a function of the sine of that angle.

Single-mode fiber - An optical fiber that supports only one mode of light propagation above the cutoff wavelength.

Small Computer Systems Interface (SCSI) - A standard for a controller bus that connects disk drives and other devices to their controllers on a computer bus. It is typically used in small systems.

Snail Mail - A derogatory term referring to the US Postal Service speed as compared to electronic mail.

SNR - Signal-to-noise ratio.

Software (computer) - Programming, especially problem-oriented.

Solenoid - An electromagnet with a moveable iron core.

Source - the light emitter, either an LED or laser diode, in a fiber-optic link.

Specific Gravity - The ratio of the weight of the solution to the weight of an equal volume of water at a specified temperature. Used as an indicator of battery state of charge.

Spectral width - A measure of the extent of a spectrum. For a source, the width of wavelengths contained in the output at one half of the wavelength of peak power. Typical spectral widths are 20 to 60 nm for an LED and 2 to 5 nm for a laser diode.

Splice, fusion or mechanical - A device that provides for a connection between two fibers, typically intended to be permanent.

Splice - An interconnection method for joining the ends of two optical fibers in a permanent or semipermanent fashion.

SRAM - Static RAM

Star coupler - A fiber-optic coupler in which power at any input port is distributed to all output ports.

Star network - A network in which all terminals are connected through a single point, such as a star coupler.

Step-index fiber - An optical fiber, either multimode or single mode, in which the core refractive index is uniform throughout so that a sharp step in refractive index occurs at the core-to-cladding interface. It usually refers to a multi-mode fiber.

STM - Synchronous Transfer Mode

STP - Shielded Twisted Pair.

Strength member - the part of a fiber-optic cable composed of Kevlar aramid yarn, steel strands, or fiberglass filaments that increase the tensile strength of the cable.

Surface emitter LED - A LED that emits light perpendicular to the semiconductor chip. Most LEDs used in datacommunications are surface emitters.

Surge Capacity - The requirement of an invertor to tolerate a momentary current surge imposed by starting ac motors or transformers.

Switch - Telephone equipment used to interconnect lines and trunks.

Switched Line - A communication link that may vary the physical path with each usage.

Systems Network Architecture (SNA) - A proprietary networking architecture used by IBM and IIBM-compatible mainframe computers. With its widespread use, SNA has become a de facto standard.

T1 - A digital carrier facility for transmitting a single DS1 digital stream over two pairs of regular copper telephone wires at 1.544 Mbps. It has come to mean any 1.544 Mbps digital stream, regardless of what transmission medium.

T2 - A digital carrier facility used to transmit a DS2 digital stream at 6.312 Mbps.

T3 - A digital carrier facility used to transmit a DS3 digital stream at 44.746 megabits per second. It has come to mean any 44.746 Mbps digital stream, regardless of what transmission medium.

T4 - A digital carrier facility used to transmit a DS4 digital stream at 273m bps.

Talkset, fiber optic - A communication device that allows conversation over unused fibers.

Tap loss - In a fiber-optic coupler, the ratio of power at the tap port to the power at the input port.

Tap port - In a coupler in which the splitting ratio between output ports is not equal, the output port containing the lesser power.

TDM - Time-division multiplexing.

Tee coupler - A three-port optical coupler.

Teflon - The Dupont brand name for HDPE resin. This material is often used to comply with smoke resistance requirements of the National Electric Code. Plenum cable is then frequently Teflon cable.

Telco - Local telephone company.

Telecommunications - The transmission of voice and/or data through a medium by means of electrical impulses that includes all aspects of transmitting information.

Telecommunications Outlet - The end of the fixed cable system in the users work area. Usually a wall or floor jack.

Termination - Preparation of the end of a fiber to allow connection to another fiber or an active device, sometimes also called "connectorization".

Test cable - A short single-fiber jumper cable with connectors on both ends used for testing. This cable must be made of fiber and connectors of a matching type to the cables to be tested.

Test kit - A kit of fiber optic instruments, typically including a power meter, source, and test accessories, used for measuring loss and power.

Test source - A laser diode or LED used to inject an optical signal into fiber for testing loss of the fiber or other components.

Thermal protection - Refers to an electrical device which has inherent protection from overheating. Typically in the form of a bimetal strip which bends when heated to a certain point. When the bimetal strip is used as a part of appliance's circuitry, the circuit will open when the bimetal bends, breaking the circuit.

Thyristor - A family of switching semiconductor devices, including SCRs, triacs, and diacs.

Time Division Multiplexing - A system of multiplexing that allocates time slices to each input channel for carrying data over an aggregate circuit.

Token Bus - A network access mechanism and topology in which all stations attached to the bus listen for a token. Stations with data to send must have the token to transmit their data. Bus access is controlled by preassigned priority algorithms. It is most used by ARCNet.

Token Passing - A network access method in which the stations circulate a token. Stations with data to send must have the token to transmit their data. See also Token Ring and Token Bus.

Token Ring - A network access method and topology in which a token is passed from station to station in sequential order. Stations wishing to send data must wait for the token before transmitting data. In a token ring, the next logical station is also the next physical station on the ring.

Topology - A standard method of connecting systems on a network.

Total internal reflection - Confinement of light into the core of a fiber by the reflection off the core-cladding boundary.

Tracking Array - A PV array that follows the daily path of the sun. This can mean one axis or two axis tracking.

Traffic - 1) Calls being sent and received over a communications network. 2) The packets that are sent on a data network.

Transceiver - Transmitter-receiver.

Transducer - A device for converting energy from one form to another, such as optical energy to electrical energy.

Transformer - A device which uses magnetic force to transfer electrical energy from one coil of wire to another. In the process, transformers can also change the voltage at which this electrical energy is transmitted.

Transmission - The electrical transfer of a signal, message or other form of data from one location to another.

Transmission Speed - The number of bits transmitted in a given period of time, usually expressed as Bits Per Second (bps).

Traveling wave tube - A UHF electron tube in which a wave travelling along a helix interacts with an electron beam traveling down the center of the helix.

Trunk - A telephone circuit that connects two switches.

Twisted Pair - A type of cable in which pairs of conductors are twisted together to produce certain electrical properties. See also shielded twisted pair and unshielded twisted pair.

Tx - Transmit

Underwriters Laboratories (UL) - A non-profit organization that was established by the insurance industry to test devices, materials and systems for safety, not satisfactory operation. It has begun to set standards. Items that pass the tests are marked UL Approved.

Uninterrupted Power Supply (UPS) - Designation of a power supply providing continuous uninterrupted service.

Uninterruptible Power Supply (UPS) - A device that provides continuous power in case the main power source fails. It includes filtering that provides a high quality AC power signal. See Standby Power System.

UNIX - A multi-user operating system developed by Bell Laboratories.

Unshielded Twisted Pair (UTP) - Cable that consists of two or more insulated conductors in which each pair of conductors are twisted around each other. There is no external protection and noise resistance comes solely from the twists. See also Category Cable.

Up Link - The connection from an earth station to a satellite.

UPS - See Uninterruptible power supply.

Usenet - The thousands of topically named newsgroups, the computers which run them, and the people who read and submit Usenet news.

Utilization equipment - Equipment which uses electricity.

UTP - See Unshielded Twisted Pair

Velocity of Propagation - See Propagation Velocity.

Virus - A program that replicates itself, usually to the detriment of the system, by incorporating itself into other programs which are shared among computer systems.

Visual fault locator - A device that couples visible light into the fiber to allow visual tracing and testing of continuity. Some are bright enough to allow finding breaks in fiber through the cable jacket.

Voice Frequency (VF) - Any of the frequencies in the band 300-3, 400 Hz which must be transmitted to reasonably reproduce the voice.

Voice Grade - An access line suitable for voice, low-speed data, or facsimile service.

Volatile Storage - Any memory that loses it contents when electrical power is removed.

Volt - The unit of measurement of electrical force. One volt will force one ampere of current to flow through a resistance of one ohm.

Voltage Drop - Voltage reduction due to wire resistance.

VSAT - Very Small Aperture Terminal. A smaller dish for satellite communication.

WAN - Wide area network

WATS - Wide Area Telephone Service.

Watt - The unit of measurement of electrical power or rate of work. One amp represents the amount of work that is done by one ampere of current at a pressure of one volt.

Waveform - Characteristic shape of an electrical current or signal. The ace output from an invertor.

Wavelength division multiplexing (WDM) - A technique of sending signals of several different wavelengths of light into the fiber simultaneously.

Wavelength - The distance between the same two points on adjacent waves; the time required for a wave to complete a single cycle.

Wavelength-division multiplexing - A transmission technique by which separate optical channels, distinguished by wavelength, are multiplexed onto an optical fiber for transmission.

WDM - Wavelength-division multiplexing.

Whips - A flexible assembly, usually of THHN conductors in flexible metal conduit with fittings, usually bringing power from a lighting outlet to a lighting fixture.

Wide Area Network (WAN) - A network that covers a large geographic area. See also Local Area Network, Metropolitan Area Network.

Wide Area Telephone Service (WATS) - A telco service that lets a customer make calls to or from telephones on designated lines, with a discounted monthly charge based upon call volume. Typically synonymous with "800" service. This is not free calling as SOMEONE must pay for every call.

Wideband - A term applied to facilities that have bandwidths greater than those needed for one channel.

Windows - A generic term for the Microsoft extension to MS-DOS that offers limited multi-tasking on Intel-based computers.

Windows '95 - Microsoft's latest version of Windows.

Windows NT - Windows New Technology. The general term for Microsoft's 32 bit operating system. It was designed to supplement or replace Windows and MS-DOS.

Wire Center - The physical structure that houses one or more central office switching systems.

Wiring Concentrator - See concentrator.

Workstation - A networked computing device with additional processing power and RAM. Often, a workstation runs an operating system such as UNIX or Windows or Apple's System 7 so several tasks can run simultaneously.

World Wide Web (WWW) - A project that merges information retrieval and hypertext to make an easy to use, powerful, global, academic information system.

Worm - A computer program that replicates and distributes itself. Worms, as opposed to viruses, are meant to spread in network environments.

WWW - See World Wide Web

WYSIWYG - What You See Is What You Get. A word processing term that indicates the screen display will match the printer output.

Xfer - Transfer

Xmodem - An eight bit, public domain error checking protocol.

Ymodem - A file transfer protocol based on CRC Xmodem. Ymodem has a 1024-byte packet size.

Zip - To compress a file or program. Usually done with PKZIP.

Zmodem - An error-correcting, full duplex, file transfer, data transmission protocol for copying files between computers. Most agree that it is faster and better than Xmodem and Ymodem.